Modern Astronomy

Startling Facts

With 1001 Questions and Answers

Modern Astronomy

Startling Facts

With 1001 Questions and Answers

C Sivaram
Senior Professor and Chairman Theory Group
Indian Institute of Astrophysics
Bengalore-560034

Kenath Arun
Lecturer
Department of Physics, Christ Junior College
Christ University Campus
Bengalore-560029

Ane Books Pvt. Ltd.

New Delhi ♦ Chennai

Modern Astronomy Startling Facts
C Sivaram and Kenath Arun

Price: ₹ 495/-
Pages: viii + 174 (Pb)
Size : Royal (6¼" × 9½")
Paper Quality (GSM) : 80

Published by

Ane Books Pvt. Ltd.

4821, Parwana Bhawan, 1st Floor, 24 Ansari Road, Darya Ganj,
New Delhi - 110 002, Tel.: +91(011) 23276843-44, Fax: +91(011) 23276863
e-mail: kapoor@anebooks.com, Website: www.anebooks.com

Branch Office :

- Avantika Niwas, 1st Floor, 19 Doraiswamy Road, T. Nagar,
Chennai - 600 017, Tel.: +91(044) 28141554, 28141209
e-mail: anebookschennai@gmail.com, rathinam@anebooks.com

ISBN : 978-93-8015-605-7

Printed at : R.K. Offset Process, New Delhi.

Preface

The year 2009 has been declared as the International Year of Astronomy. A lot of activity is on to stimulate interest in astronomy among students, general public and others. Many lectures, talks, movies, quizzes, etc. are being planned and some are already underway. Towards this we feel that an Astronomy Quiz book, along with introductory notes on a wide variety of topics in astronomy, would be very timely. We have, in this book, a thousand interesting trivia related to all aspects of astronomy.

Also included are introduction to wide range of topics on modern astronomy, including planets, stars, space probes, astronomers: facts and discoveries, historical facts, observatories, etc.

We hope this book will be of good interest to both general public as well as to students interested in astronomy and its many interesting fields

— Authors

Preface

The year 2009 has been declared as the International Year of Astronomy. A lot of activity is on to stimulate interest in astronomy among students, general public and others. Many lectures, talks, movies, quizzes, etc. are being planned and some are already underway. Towards this we feel that an Astronomy Quiz book, along with introductory notes on a wide variety of topics in astronomy, would be very timely. We have, in this book, a thousand interesting trivia related to all aspects of astronomy.

Also included are introduction to wide range of topics on modern astronomy, including planets, stars, space probes, astronomers: facts and discoveries, historical facts, observatories, etc.

We hope this book will be of good interest to both general public as well as to students interested in astronomy and its many interesting fields

— Authors

Contents

International Year of Astronomy, 2009

The year 2009 is being recognised as the international year of astronomy (IYA). It marks the four hundredth anniversary of the historical occasion in the year 1609 when Galileo Galilei used the then newly invented telescope to observe astronomical object and soon after publishing his discoveries in his work Sidereus Nuncius.

Especially his discovery of the four moons (satellites) orbiting the giant planet Jupiter opened up the astronomical frontiers while providing support to the Copernican picture of the earth being no longer the centre of the universe around which everything revolved! Today we know there are more than one hundred and sixty moons orbiting the giant planets with Jupiter and Saturn having several dozens of satellites each.

The past decade has witnessed the discovery of well over three hundred exoplanets orbiting nearby stars and hundreds of smaller Trans-Neptunian Objects orbiting in the outskirts of our own solar system.

Galileo just had a puny three inch telescope. Today we have giant telescopes well over ten metres across with plans for thirty metre and hundred metre telescopes in the near future! The Hubble Space Telescope has a two metre aperture and has been orbiting the earth for nearly two decades.

The light gathering power of our giant telescopes is well over a hundred thousand times that of Galileo's tiny instrument and has enabled us to glimpse galaxies and other objects more than ten billion light years away. These objects formed nearly nine billion years before the sun and the solar system were born! (A billion is thousand million)

Today's telescopes are accompanied by paraphernalia of sophisticated instruments like photometers, spectrometers, etc. enabling astronomers to detect exotic elements like uranium, thorium (and even unstable elements like technetium, which is no longer present on earth), in distant stars and also to monitor motions and positions of these objects to high precision.

We know the distance to the moon to within a fraction of a centimetre and can measure the slowing down of the earth's rotation (about a millisecond per century) with the help of atomic clocks. Astronomical knowledge has enabled us to explore all the planets (and even comets and asteroids) in our solar system by sending spacecraft to these objects to land, orbit or fly past them.

Again astronomy today is no longer confined to the optical part of the spectrum that is the electromagnetic part of the spectrum to whose wavelengths our eyes are sensitive to. Astronomers have literally realised that there is far far more to the universe than just what meets our eyes! They have opened up for us to behold visions of a universe vastly vaster than ever imagined with images of celestial objects in solitary splendour scintillating over all octaves of the spectrum as well as spectacular phenomena pervading over stupendous scales of distance, time and energy.

We have distant objects like quasars emitting energy in X-rays itself, in one second, a hundred trillion times the sun's luminosity. The sun itself emits in one second the amount of energy that mankind will consume (at the present global power production of ten trillion watts) in a hundred million years. Advances in nuclear physics have enabled us to understand the precise sequence of nuclear interactions that powers the sun's vast energy output.

In recent years this has involved detecting subatomic particles called neutrinos, which are copiously produced in these nuclear interactions and pass right through the sun. A hundred trillion of these ghostly solar neutrinos pass through our bodies every second without interaction! Yet very large detectors deep underground have detected these neutrinos, opening up a new branch of astronomy, neutrino astronomy! Ironically to peer deep inside as to see that is happening in the solar interior we have to have laboratories deep underground!

The stars range in size from a billion kilometres (red giants) to ten kilometres (neutron stars, the size of a city!). They have densities from being much more rarefied than air to that of a billion tons per cubic centimetre. There are stars emitting several million times what the sun emits! While nearest star Proxima Centauri is twenty thousand times fainter than the sun and not visible to the naked eye.

Yet the realisation that all these types of stellar objects are the various stages of evolution in the lives of stars has enabled us to get a detailed picture of how stars live and die! The sun will end up a white dwarf. Massive stars will explode as supernovae which at their peak emit ten billion times the sun's luminosity and are visible billions of light years away! The most

massive ones have cores that collapse into so called black holes; their gravity field being so strong that even light is trapped.

Several galaxies including our own host supermassive black holes in their core. Quasars are powered by such exotic objects which accrete the surrounding matter. Astronomers can study all these energetic objects in all wavelengths, X-rays, gamma rays, ultraviolet, infrared, radio waves, etc. Radioastronomy has led to the precise monitoring of pulsars (spinning neutron stars), some of them slowing down so precisely that they keep time better than atomic clocks!

A big discovery of radioastronomy has been that of cosmic microwave background, the left over all pervading remnant radiation from the hot dense early phase (the so called big bang) of our universe. This led to two Nobel Prizes. Pulsar radioastronomy also led to two Nobel Prizes. We also have cosmic X-rays and gamma rays background radiation around us.

In February 1987, a massive star exploded as a supernova in our neighbouring galaxy, the Large Magellanic Cloud. Neutrinos emitted from its collapsing core were detected on earth as expected from theory. Every second, a supernova is exploding somewhere in the universe.

We are witnessing puzzling very energetic events like gamma ray bursts (emitting in one second high energy gamma rays equal to the entire radiative output of the sun in ten billion years!), colliding galaxies etc.

The most energetic events like collisions of supermassive black holes, tidal disruptions of neutron stars, etc. would in future be monitored by gravitational wave astronomy. We already have the LIGO detector installed over three continents. The Auger detector covering several thousand square kilometres is detecting the highest energy cosmic rays (each tiny particle packing the energy of a falling brick!). While the ice cube detector in Antarctica detects highest energy neutrinos from distant exotic objects.

Last but not the least there is excitement about astrobiology and detection of life, especially possible intelligent life, on the innumerable worlds populating the universe. Powerful radio beacons and laser beams from advanced civilisations are already being seriously searched for using the largest telescopes. 2009 also marks the sesquicentury of Darwin's epochal work 'Origin of Species' which revolutionised biology and the bicentenary of Darwin's birth. So it would be appropriate to have 2009 also as a year for astrobiology as part of the year of astronomy.

In short, astronomy, one of the oldest of sciences is in for a big leap in the coming decades. Let us not forget that among the most recent enigmatic discoveries have been that dark matter and dark energy constitute about ninety five per cent of what the universe is made up of and we are completely ignorant at present about what they really are!

Several space probes and future generation space telescopes will try to uncover this mystery, complimenting several sophisticated laboratory dark matter searches on earth. During IYA let us ponder over these enigmas in all humility.

□□□

1

The Solar System

The Solar System consists of the Sun and those celestial objects bound to it by gravity. These objects are the eight planets, their 166 known moons, five dwarf planets, and billions of small bodies. The small bodies include asteroids, icy Kuiper belt objects, comets, meteoroids, and interplanetary dust.

The charted regions of the Solar System are the Sun, four terrestrial inner planets, the asteroid belt, four gas giant outer planets, the Kuiper belt. The Oort cloud exists at a distance roughly a thousand times beyond the region of Kuiper belt.

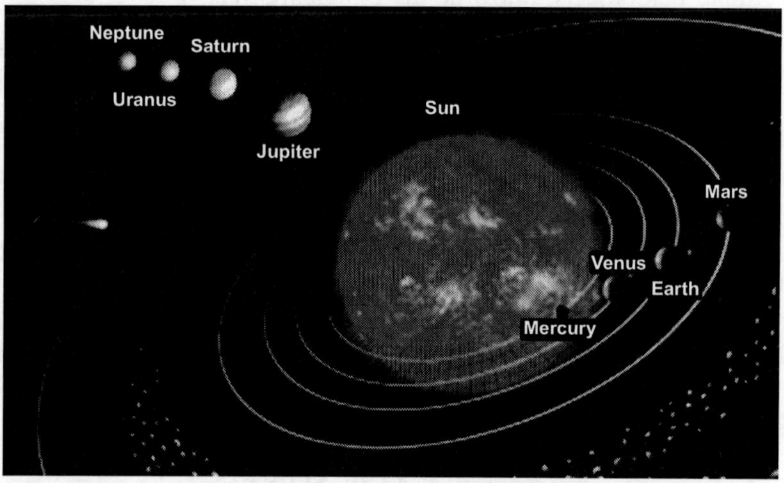

Fig. 1.1 : The solar system

In order of their distances from the Sun, the eight planets are:

1. Mercury,
2. Venus,
3. Earth,
4. Mars,
5. Jupiter,
6. Saturn,
7. Uranus,
8. Neptune.

The Solar System is dominated by the Sun, a main sequence star that contains 99.86 per cent of the system's known mass and dominates it gravitationally. Jupiter and Saturn, the Sun's two largest orbiting bodies, account for more than 90 per cent of the system's remaining mass.

Most large objects in orbit around the Sun lie near the plane of Earth's orbit, known as the ecliptic. The planets are very close to the ecliptic while comets and Kuiper belt objects are usually at significantly greater angles to it. All of the planets and most other objects also orbit with the Sun's rotation (counter-clockwise, as viewed from above the Sun's North Pole). There are exceptions, such as Halley's Comet.

Kepler's laws of planetary motion describe the orbits of objects about the Sun. According to Kepler's laws, each object travels along an ellipse with the Sun at one focus. Objects closer to the Sun (with smaller semi-major axes) have shorter years.

On an elliptical orbit, a body's distance from the Sun varies over the course of its year. A body's closest approach to the Sun is called its *perihelion*, while it's most distant point from the Sun is called its *aphelion*.

Each body moves fastest at its perihelion and slowest at its aphelion. The orbits of the planets are nearly circular, but many comets, asteroids and Kuiper belt objects follow highly elliptical orbits.

In reality, with a few exceptions, the farther a planet or belt is from the Sun, the larger the distance between it and the previous orbit. For example, Venus is approximately 0.33 astronomical units (AU) farther out than Mercury, while Saturn is 4.3 AU out from Jupiter, and Neptune lies 10.5 AU out from Uranus. Attempts have been made to determine a correlation between these orbital distances (like Titius-Bode law).

Most of the planets in the Solar System possess secondary systems of their own. Many are in turn orbited by planetary objects called natural satellites, or moons, some of which are larger than planets. Most of the largest natural satellites are in synchronous orbit, with one face permanently turned toward their parent. The four largest planets also possess planetary rings, thin bands of tiny particles that orbit them in unison.

1.1 THE SUN

The Sun is the Solar System's parent star. Its large mass gives it an interior density high enough to sustain nuclear fusion, which releases enormous amounts of energy, mostly radiated into space as electromagnetic radiation such as visible light.

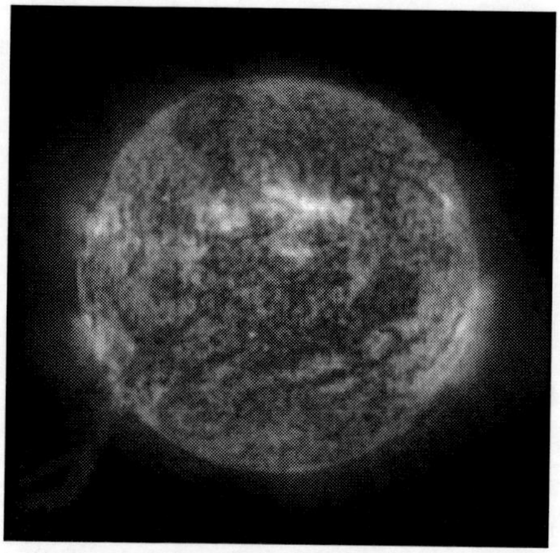

Fig. 1.2:
The sun (with flares and prominences)

The Sun is classified as a moderately large yellow dwarf. Stars are classified by the Hertzsprung-Russell diagram, shown in Fig.1.3 a graph which plots the brightness of stars against their surface temperatures. Generally, hotter stars are brighter. Stars following this pattern are said to be on the main sequence; the Sun lies right in the middle of it.

The Sun is a population I star; it was born in the later stages of the universe's evolut-ion. It contains more elements heavier than hydrogen and helium ("metals" in astronomical parlance) than older population II stars. Elements heavier

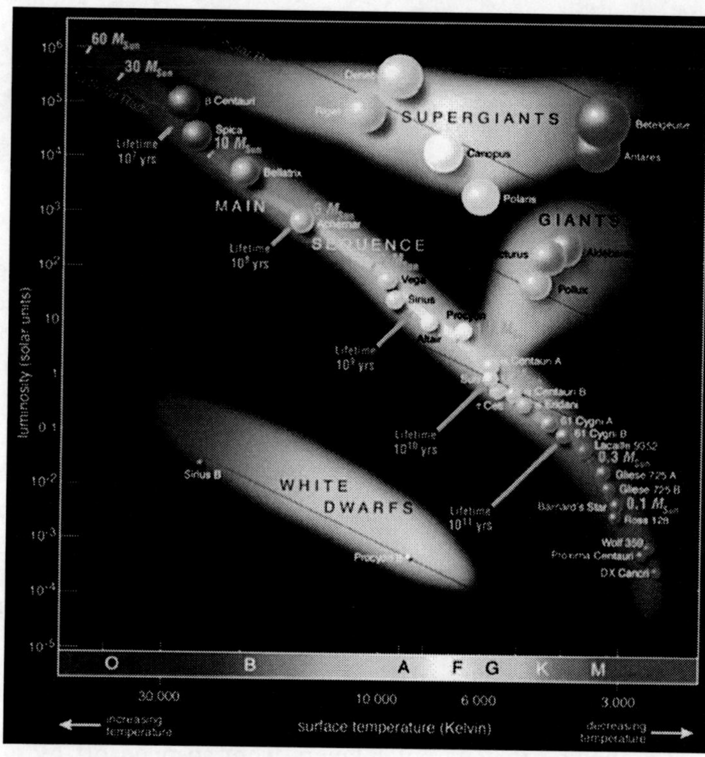

Fig. 1.3: H-R diagram showing the luminosity-temperature relation

than hydrogen and helium were formed in the cores of ancient and exploding stars, so the first generation of stars had to die before the universe could be enriched with these atoms. The oldest stars contain few metals, while stars born later have more. This high metallicity is thought to have been crucial to the Sun's developing a planetary system, because planets form from accretion of metals.

1.2 THE TERRESTRIAL PLANETS

The four inner or terrestrial planets have dense, rocky compositions, few or no moons, and no ring systems. They are composed largely of minerals with high melting points, such as the silicates which form their crusts and mantles, and metals such as iron and nickel, which form their cores.

Fig. 1.4 : Relative sizes of the terrestrial planets

Three of the four inner planets (Venus, Earth and Mars) have substantial atmospheres; all have impact craters and tectonic surface features such as rift valleys and volcanoes. The term *inner planet* should not be confused with *inferior planet,* which designates those planets which are closer to the Sun than Earth is (*i.e.,* Mercury and Venus).

1.2.1 Mercury

Mercury (0.4 AU) is the closest planet to the Sun and the smallest planet (0.055 Earth masses). Mercury has no natural satellites, and it's only known geological features besides impact craters are lobed ridges or rupes, probably produced by a period of contraction early in its history. Mercury's almost negligible atmosphere consists of atoms blasted off its surface by the solar wind. Its relatively large iron core and thin mantle have not yet been adequately explained.

Hypotheses include that its outer layers were stripped off by a giant impact, and that it was prevented from fully accreting by the young Sun's energy.

Fig. 1.5: Mercury

1.2.2 Venus

Venus (0.7 AU) is close in size to Earth, (0.815 Earth masses) and like Earth, has a thick silicate mantle around an iron core, a substantial atmosphere and evidence of internal geological activity.

Fig. 1.6: Venus

However, it is much drier than Earth and its atmosphere is ninety times as dense. Venus has no natural satellites. It is the hottest planet, with surface temperatures over 400°C, most likely due to the amount of greenhouse gases in the atmosphere.

No definitive evidence of current geological activity has been detected on Venus, but it has no magnetic field that would prevent depletion of its substantial atmosphere, which suggests that its atmosphere is regularly replenished by volcanic eruptions.

1.2.3 Earth

Earth (1 AU) is the largest and densest of the inner planets, the only one known to have current geological activity, and the only planet known to have life. Its liquid hydrosphere is unique among the terrestrial planets, and it is also the only planet where plate tectonics has been observed. Earth's atmosphere is radically different from those of the other planets, having been altered by the presence of life to contain 21% free oxygen. It has one natural satellite, the Moon, the only large satellite of a terrestrial planet in the Solar System.

1.2.4 Mars

Fig.1.7 : Mars

Mars (1.5 AU) is smaller than Earth and Venus (0.107 Earth masses). It possesses a tenuous atmosphere of mostly carbon dioxide. Its surface, peppered with vast volcanoes such as Olympus Mons and rift valleys such as Valles Marineris, shows geological activity that may have persisted until very recently. Its red colour comes from rust in its iron-rich soil. Mars has two tiny natural satellites (Deimos and Phobos) thought to be captured asteroids.

1.3 ASTEROID BELT

Asteroids are mostly small Solar System bodies composed mainly of rocky and metallic non-volatile minerals. The main asteroid belt occupies the orbit between Mars and Jupiter, between 2.3 and 3.3 AU from the Sun. It is thought to be remnants from the Solar System's formation that failed to coalesce because of the gravitational interference of Jupiter. Asteroids range in size from hundreds of kilometres across to microscopic.

All asteroids save the largest, Ceres, are classified as small Solar System bodies, but some asteroids such as Vesta and Hygieia may be reclassified as dwarf planets if they are shown to have achieved hydrostatic equilibrium. The asteroid belt contains tens of thousands, possibly millions, of objects over one kilometre in diameter. Despite this, the total mass of the main belt is unlikely to be more than one-thousandth of that of the Earth. The main belt is very sparsely populated; spacecraft routinely pass through without incident. Asteroids with diameters between 10 and 10^{-4}m are called meteoroids.

Asteroids in the main belt are divided into asteroid groups and families based on their orbital characteristics. Asteroid moons are asteroids that orbit larger asteroids. They are not as clearly distinguished as planetary moons, sometimes being almost as large as their partners. The asteroid belt also

Fig. 1.8 : Asteroid belt

contains main-belt comets which may have been the source of Earth's water. Trojan asteroids are located in either of Jupiter's L_4 or L_5 points (gravitationally stable regions leading and trailing a planet in its orbit); the term "Trojan" is also used for small bodies in any other planetary or satellite Lagrange point. Hilda asteroids are in a 2:3 resonance with Jupiter; that is, they go around the Sun three times for every two Jupiter orbits. The inner Solar System is also dusted with rogue asteroids, many of which cross the orbits of the inner planets.

1.4 THE GAS GIANTS

The four outer planets, or gas giants (or Jovian planets), collectively make up 99 per cent of the mass known to orbit the Sun. Jupiter and Saturn consist overwhelmingly of hydrogen and helium; Uranus and Neptune possess a greater proportion of ices in their makeup. Some astronomers suggest they belong in their own category, "ice giants." All four gas giants have rings, although only Saturn's ring system is easily observed from Earth.

Fig. 1.9: The Jovian planets

1.4.1 Jupiter

Jupiter (5.2 AU), at 318 Earth masses, masses 2.5 times all the other planets put together. It is composed largely of hydrogen and helium. Jupiter's strong internal heat creates a number of semi-permanent features in its atmosphere, such as cloud bands and the Great Red Spot. Jupiter has sixty-three known satellites. The four largest, Ganymede, Callisto, Io, and Europa, show similarities to the terrestrial planets, such as volcanism.

Fig.1.10: Jupiter

1.4.2 Saturn

Saturn (9.5 AU), distinguished by its extensive ring system, has similarities to Jupiter, such as its atmospheric composition. Saturn is far less massive,

Fig. 1.11: Saturn

being only 95 Earth masses. Saturn has sixty known satellites (and three unconfirmed); two of which, Titan and Enceladus, show signs of geological activity, though they are largely made of ice. Titan is larger than Mercury and the only satellite in the Solar System with a substantial atmosphere.

1.4.3 Uranus

Fig. 1.12: Uranus

Uranus (19.6 AU), at 14 Earth masses, is the lightest of the outer planets.Uniquely among the planets, it orbits the Sun on its side; its axial tilt is over ninety degrees to the ecliptic. It has a much colder core than the other gas giants, and radiates very little heat into space. Uranus has twenty-seven known satellites, the largest ones being Titania, Oberon, Umbriel, Ariel and Miranda.

1.4.4 Neptune

Neptune (30 AU), though slightly smaller than Uranus, is more massive (equivalent to 17 Earths) and therefore more dense. It radiates more internal heat, but not as much as Jupiter or Saturn. Neptune has thirteen known satellites. The largest, Triton, is geologically active, with geysers of liquid nitrogen. Triton is the only large satellite with a retrograde orbit. Neptune is

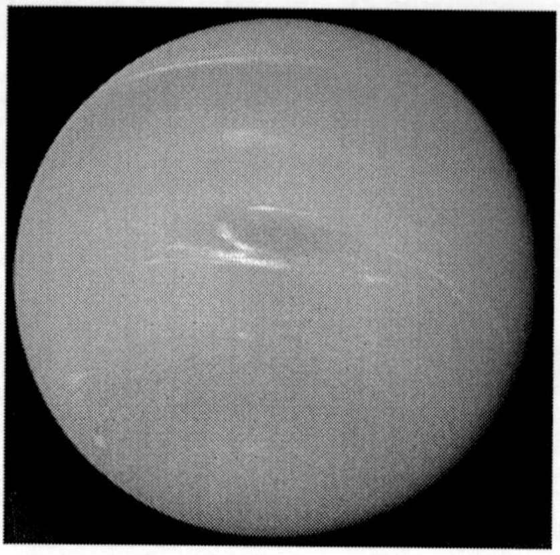

Fig. 1.13: Neptune

accompanied in its orbit by a number of minor planets, termed Neptune Trojans that are in 1:1 resonance with it.

1.5 COMETS

Comets are small Solar System bodies, usually only a few kilometres across, composed largely of volatile ices. They have highly eccentric orbits, generally a perihelion within the orbits of the inner planets and an aphelion far beyond Pluto. When a comet enters the inner Solar System, its proximity to the Sun causes its icy surface to sublimate and ionise, creating a coma: a long tail of gas and dust often visible to the naked eye.

Fig. 1.14: Comet with the two tails

Short-period comets have orbits lasting less than two hundred years. Long-period comets have orbits lasting thousands of years. Short-period comets are believed to originate in the Kuiper belt, while long-period comets, such as Hale-Bopp, are believed to originate in the Oort cloud.

Many comet groups, such as the Kreutz Sungrazers, formed from the break-up of a single parent. Some comets with hyperbolic orbits may originate outside the Solar System, but determining their precise orbits is difficult. Old comets that have had most of their volatiles driven out by solar warming are often categorised as asteroids.

The area beyond Neptune, or the "trans-Neptunian region", is still largely unexplored. It appears to consist overwhelmingly of small worlds (the largest having a diameter only a fifth that of the Earth and a mass far smaller than that of the Moon) composed mainly of rock and ice. This region is sometimes known as the "outer Solar System", though others use that term to mean the region beyond the asteroid belt.

1.6 THE KUIPER BELT

The Kuiper belt, the region's first formation, is a great ring of debris similar to the asteroid belt, but composed mainly of ice. It extends between 30 and 50 AU from the Sun. It is composed mainly of small Solar System bodies, but many of the largest Kuiper belt objects, such as Quaoar, Varuna, and Orcus, may be reclassified as dwarf planets.

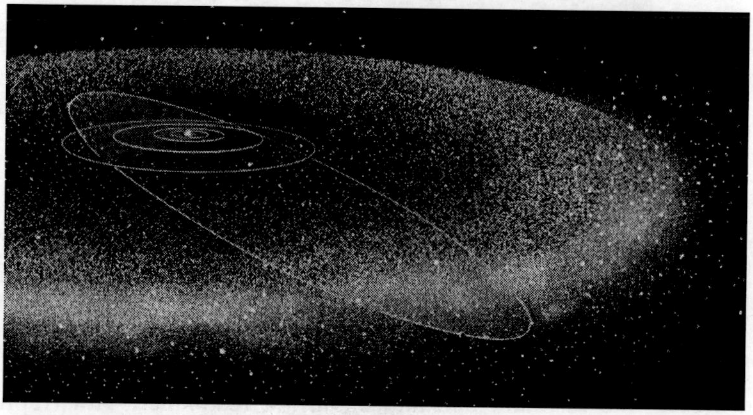

Fig. 1.15: Kuiper Belt objects (well beyond the orbit of Neptune)

There are estimated to be over 100,000 Kuiper belt objects with a diameter greater than 50 km, but the total mass of the Kuiper belt is thought to be only a tenth or even a hundredth the mass of the Earth. Many Kuiper belt objects have multiple satellites, and most have orbits that take them outside the plane of the ecliptic.

The Kuiper belt can be roughly divided into the "classical" belt and the resonances. Resonances are orbits linked to that of Neptune (*e.g.*, twice for every three Neptune orbits, or once for every two). The first resonance actually begins within the orbit of Neptune itself.

The classical belt consists of objects having no resonance with Neptune, and extends from roughly 39.4 AU to 47.7 AU. Members of the classical Kuiper belt are classified as cubewanos, after the first of their kind to be discovered, (15760) 1992 QB_1.

1.6.1 Pluto

Pluto (39 AU average), a dwarf planet, is the largest known object in the Kuiper belt. When discovered in 1930, it was considered to be the ninth planet; this changed in 2006 with the adoption of a formal definition of planet.

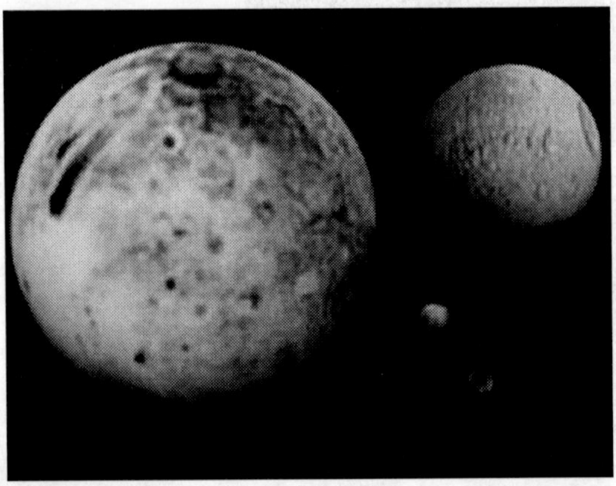

Fig. 1.16: Pluto with Charon, Nix and Hydra (Plutoids)

Pluto has a relatively eccentric orbit inclined 17 degrees to the ecliptic plane and ranging from 29.7 AU from the Sun at perihelion (within the orbit of Neptune) to 49.5 AU at aphelion. Pluto and its three known moons: Both Pluto and Charon orbit a barycenter of gravity above their surfaces, making Pluto-Charon a binary system. Two much smaller moons, Nix and Hydra, orbit Pluto and Charon.

Pluto lies in the resonant belt and has a 3:2 resonance with Neptune, meaning that Pluto orbits twice round the Sun for every three Neptunian orbits. Kuiper belt objects whose orbits share this resonance are called plutinos.

1.7 THE OORT CLOUD

The Oort cloud is a great mass of up to a trillion icy objects that is believed to be the source for all long-period comets and to surround the Solar System at roughly 50,000 AU (around 1 light-year (LY)), and possibly to as far as 100,000 AU (1.87 LY). It is believed to be composed of comets which were ejected from the inner Solar System by gravitational interactions with the outer planets.

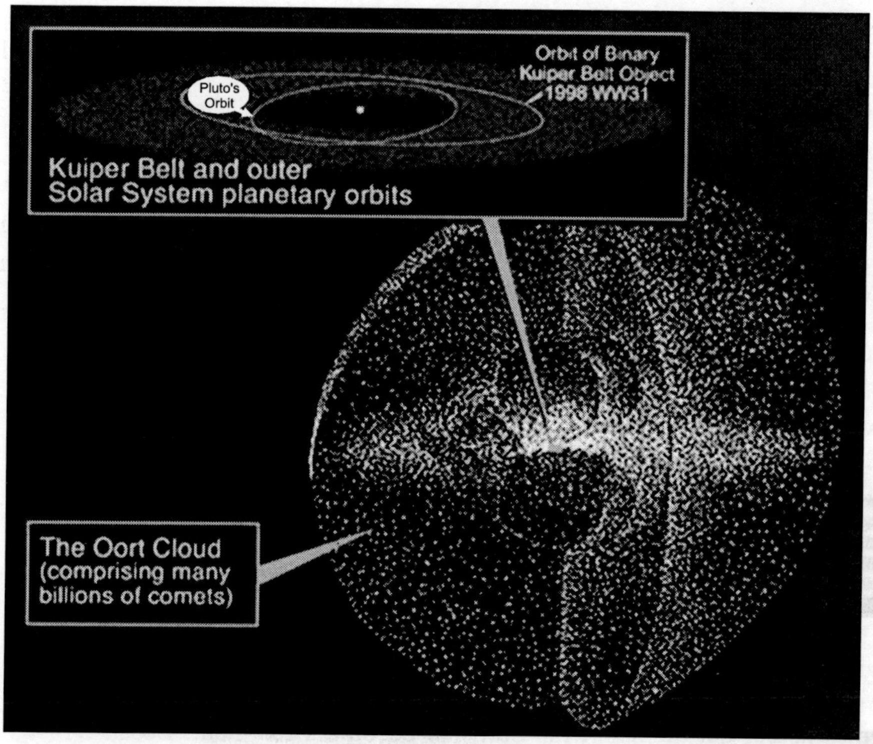

Fig. 1.17: Oort cloud, extending much beyond the inner solar system

Oort cloud objects move very slowly, and can be perturbed by infrequent events such as collisions, the gravitational effects of a passing star, or the galactic tide, the tidal force exerted by the Milky Way.

90377 Sedna (525.86 AU average) is a large, reddish Pluto-like object with a gigantic, highly elliptical orbit that takes it from about 76 AU at perihelion to 928 AU at aphelion and takes 12,050 years to complete.

1.8 EVOLUTION OF SOLAR SYSTEM

The Solar System formed from the gravitational collapse of a giant molecular cloud 4.6 billion years ago. This initial cloud was likely several light-years across and probably birthed several stars.

As the region that would become the Solar System, known as the pre-solar nebula, collapsed, conservation of angular momentum made it rotate faster. The centre, where most of the mass collected, became increasingly hotter than the surrounding disc. As the contracting nebula rotated, it began to flatten into a spinning protoplanetary disc with a diameter of roughly 200 AU and a hot, dense protostar at the centre.

At this point in its evolution, the Sun is believed to have been a T Tauri star. Studies of T Tauri stars show that they are often accompanied by discs of pre-planetary matter with masses of 0.001–0.1 solar masses, with the vast majority of the mass of the nebula in the star itself. The planets formed by accretion from this disk. Within 50 million years, the pressure and density of hydrogen in the centre of the protostar became great enough for it to begin thermonuclear fusion.

The temperature, reaction rate, pressure, and density increased until hydrostatic equilibrium was achieved, with the thermal energy countering the force of gravitational contraction. At this point the Sun became a full-fledged main sequence star.

The Solar System as we know it today will last until the Sun begins its evolution off of the main sequence of the Hertzsprung-Russell diagram [Ref: Fig.1.3]. As the Sun burns through its supply of hydrogen fuel, the energy output supporting the core tends to decrease, causing it to collapse in on itself. This increase in pressure heats the core, so it burns even faster. As a result, the Sun is growing brighter at a rate of roughly ten per cent every 1.1 billion years.

Around 5.4 billion years from now, the hydrogen in the core of the Sun will have been entirely converted to helium, ending the main sequence phase. At this time, the outer layers of the Sun will expand to roughly up to 260 times its current diameter; the Sun will become a red giant. Because of its vastly increased surface area, the surface of the Sun will be considerably cooler than it is on the main sequence (2600 K at the coolest).

Eventually, the Sun's outer layers will fall away, leaving a white dwarf, an extraordinarily dense object, and half the original mass of the Sun but only the size of the Earth. The ejected outer layers will form what is known as a planetary nebula, returning some of the material that formed the Sun to the interstellar medium.

□□□

2

Stellar Evolution

Stellar evolution begins with the gravitational collapse of a giant molecular cloud (GMC). Typical GMCs are roughly 100 light-years (9.5×10^{14} km) across and contain up to 6,000,000 solar masses (1.2×10^{37} kg). As it collapses, a GMC breaks into smaller and smaller pieces. In each of these fragments, the collapsing gas releases gravitational potential energy as heat. As its temperature and pressure increase, a fragment condenses into a rotating sphere of super-hot gas known as a protostar.

Protostars with masses less than roughly 0.08 solar mass (1.6×10^{29} kg) never reach temperatures high enough for nuclear fusion of hydrogen to begin. These are known as brown dwarfs. Brown dwarfs heavier than 13 Jupiter masses (2.5×10^{28} kg) do fuse deuterium, and some astronomers prefer to call only these objects brown dwarfs, classifying anything larger than a planet but smaller than this a sub-stellar object. Both types, deuterium-burning or not, shine dimly and die away slowly, cooling gradually over hundreds of millions of years.

For a more massive protostar, the core temperature will eventually reach 10 mega Kelvin, initiating the proton-proton chain reaction and allowing hydrogen to fuse, first to deuterium and then to helium. In stars of slightly over one solar mass (2.0×10^{30} kg), the CNO cycle contributes a considerable portion of the energy generation. The onset of nuclear fusion leads relatively quickly to a hydrostatic equilibrium in which energy released by the core exerts a "radiation pressure" balancing the weight of the star's matter, preventing further gravitational collapse. The star thus evolves rapidly to a stable state, beginning the main sequence phase of its evolution.

A new star will fall at a specific point on the main sequence of the Hertzsprung-Russell diagram, with the main sequence spectral type depending upon the mass of the star. Small, relatively cold, low mass red dwarfs burn hydrogen slowly and will remain on the main sequence for hundreds of billions of years, while massive hot supergiants will leave the main sequence after just a few million years. A mid-sized star like the Sun will remain on the main sequence for about 10 billion years. The Sun is thought to be in the middle of its lifespan; thus, it is on the main sequence.

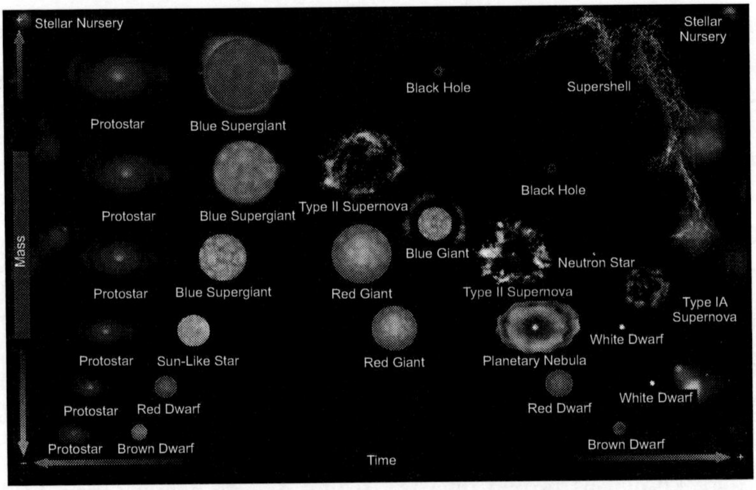

Fig 2.1: Different stages of stellar evolution

After millions to billions of years, depending on the initial mass of the star, the continuous fusion of hydrogen into helium will cause a build-up of helium in the core. Larger and hotter stars produce helium more rapidly than cooler and less massive ones.

The accumulation of helium, which is denser than hydrogen, in the core causes gravitational self-compression and a gradual increase in the rate of fusion. Higher temperatures must be attained to resist this increase in gravitational compression and to maintain a steady state.

Eventually, the core exhausts its supply of hydrogen, and without the outward pressure generated by the fusion of hydrogen to counteract the force of gravity, it contracts until either electron degeneracy becomes sufficient to oppose gravity, or the core becomes hot enough (around 100 mega Kelvin) for helium fusion to begin. Which of these happens first depends upon the star's mass.

2.1 LOW-MASS STARS

A star of less than about 0.5 solar mass will never be able to fuse helium even after the core ceases hydrogen fusion. There simply is not a stellar envelope massive enough to bear down enough pressure on the core. These are the red dwarfs, such as Proxima Centauri, some of which will live thousands of times longer than the Sun. Recent astrophysical models suggest that red dwarfs of 0.1 solar masses may stay on the main sequence for almost six trillion years, and take several hundred billion more to slowly collapse into a white dwarf.

If a star's core becomes stagnant (as is thought will be the case for the Sun), it will still be surrounded by layers of hydrogen which the star may subsequently draw upon.

However, if the star is fully convective (as thought to be the case for the lowest-mass stars), it will not have such surrounding layers. If it does, it will develop into a red giant as described for mid-sized stars below, but never fuse helium as they do; otherwise, it will simply contract until electron degeneracy pressure halts its collapse, thus directly turning into a white dwarf.

2.2 MID-SIZED STARS

In either case, the accelerated fusion in the hydrogen-containing layer immediately over the core causes the star to expand. Since this lifts the outer layers away from the core, thus reducing the gravitational pull on them, they expand faster than the energy production increases, thus causing them to cool, and thus causing the star to become redder than when it was on the main sequence. Such stars are known as red giants.

According to the Hertzsprung-Russell diagram, a red giant is a large non-main sequence star of stellar classification K or M. Examples include Aldebaran in the constellation Taurus and Arcturus in the constellation of Bootes.

A star of up to a few solar masses will develop a helium core supported by electron degeneracy pressure, surrounded by layers which still contain hydrogen. Its gravity compresses the hydrogen in the layer immediately above it, thus causing it to fuse faster than hydrogen would fuse in a main-sequence star of the same mass. This in turn causes the star to become more luminous (from 1,000 – 10,000 times brighter) and expand; the degree of expansion outstrips the increase in luminosity, thus causing the effective temperature to decrease.

The expanding outer layers of the star are convective, with the material being mixed by turbulence from near the fusing regions up to the surface of the star. For all but the lowest-mass stars, the fused material has remained deep in the stellar interior prior to this point, so the convecting envelope makes fusion products visible at the star's surface for the first time.

At this stage of evolution, the results are subtle, with the largest effects, alterations to the isotopes of hydrogen and helium, being unobservable. The effects of the CNO cycle appear at the surface, with lower $^{12}C/^{13}C$ ratios and altered proportions of carbon and nitrogen. These are detectable with spectroscopy, and have been measured for many evolved stars.

As the hydrogen around the core is consumed, the core absorbs the resulting helium, causing it to contract further, which in turn causes the remaining hydrogen to fuse even faster. This eventually leads to ignition of helium fusion (which includes the triple-alpha process) in the core. In stars of more than approximately 0.5 solar masses, electron degeneracy pressure may delay helium fusion for millions or tens of millions of years; in more massive stars, the combined weight of the helium core and the overlying layers means that such pressure is not sufficient to delay the process significantly.

When the temperature and pressure in the core become sufficient to ignite helium fusion in the core, a helium flash will occur if the core is largely supported by electron degeneracy pressure; in more massive stars, whose core is not overwhelmingly supported by electron degeneracy pressure, the ignition of helium fusion occurs relatively quietly. Even if a helium flash occurs, the time of very rapid energy release (on the order of 10^8 Suns) is brief, so that the visible outer layers of the star are relatively undisturbed.

The energy released by helium fusion causes the core to expand, so that hydrogen fusion in the overlying layers slows, and thus total energy generation decreases. Therefore, the star contracts, although not all the way to the main sequence; it thus migrates to the horizontal branch on the HR-diagram, gradually shrinking in radius and increasing its surface temperature.

After the star has consumed the helium at the core, fusion continues in a shell around a hot core of carbon and oxygen. The star follows the Asymptotic Giant Branch on the HR-diagram, paralleling the original red giant evolution, but with even faster energy generation (which thus lasts for a shorter time).

Changes in the energy output cause the star to change in size and temperature for certain periods. The energy output itself is shifted to lower frequency emission. This is accompanied by increased mass loss through powerful stellar winds and violent pulsations. Stars in this phase of life are called Late type stars, OH-IR stars or Mira-type stars, depending on their exact characteristics.

The expelled gas is relatively rich in heavy elements created within the star, and may be particularly oxygen or carbon enriched depending on the type of the star. The gas builds up in an expanding shell called a circumstellar envelope and cools as it moves away from the star, allowing dust particles and molecules to form. With the high infrared energy input from the central star ideal conditions are formed in these circumstellar envelopes for maser excitation.

Helium burning reactions are extremely sensitive to temperature, which causes great instability. Huge pulsations build up, which eventually give the outer layers of the star enough kinetic energy to be ejected, potentially forming a planetary nebula. At the centre of the nebula remains the core of the star, which cools down to become a small but dense white dwarf.

2.3 MASSIVE STARS

In massive stars, the core is already large enough at the onset of hydrogen shell burning that helium ignition will occur before electron degeneracy pressure has a chance to become prevalent. Thus, when these stars expand and cool, they do not brighten as much as lower mass stars; however, they were much brighter than lower mass stars to begin with, and are thus still brighter than the red giants formed from less massive stars. These stars are known as red supergiants.

Extremely massive stars (more than approximately 40 solar masses), which are very luminous and thus have very rapid stellar winds, lose mass so rapidly due to radiation pressure that they tend to strip off their own envelopes before they can expand to become red supergiants, and thus retain extremely high surface temperatures (and blue-white colour) from their main sequence time onwards.

Stars cannot be more than about 120 solar masses because the outer layers would be expelled by the extreme radiation. Although lower mass stars normally do not burn off their outer layers so rapidly, they can likewise avoid becoming red giants or red supergiants if they are in binary systems close enough so that the companion star strips off the envelope as it expands, or if they rotate rapidly enough so that convection extends all the way from the core to the surface, resulting in the absence of a separate core and envelope due to thorough mixing.

Core grows hotter and denser as it gains material from fusion of hydrogen at the base of the envelope. In a massive star, electron degeneracy pressure is insufficient to halt collapse by itself, so as each major element is consumed in the centre, progressively heavier elements ignite, temporarily halting collapse.

If the core of the star is not too massive (less than approximately 1.4 solar masses, taking into account mass loss that has occurred by this time), it may then form a white dwarf (possibly surrounded by a planetary nebula) as described above for less massive stars, with the difference that the white dwarf is composed chiefly of oxygen, neon, and magnesium.

Above a certain mass (estimated at approximately 2.5 solar masses, within a star originally of around 10 solar masses), the core will reach the temperature (approximately 1.1 giga Kelvin) at which neon partially breaks down to form oxygen and helium, the latter of which immediately fuses with some of the remaining neon to form magnesium; then oxygen fuses to form sulphur, silicon, and smaller amounts of other elements.

Finally, the temperature gets high enough that any nucleus can be partially broken down, most commonly releasing an alpha particle (helium nucleus) which immediately fuses with another nucleus, so that several nuclei are effectively rearranged into a smaller number of heavier nuclei, with net release of energy because the addition of fragments to nuclei exceeds the energy required to break them off the parent nuclei.

A star with a core mass too great to form a white dwarf but insufficient to achieve sustained conversion of neon to oxygen and magnesium will undergo core collapse (due to electron capture, as described above) before achieving fusion of the heavier elements. Both heating and cooling caused by electron capture onto minor constituent elements (such as aluminium and sodium) prior to collapse may have a significant impact on total energy generation within the star shortly before collapse. This may produce a noticeable effect on the abundance of elements and isotopes ejected in the subsequent supernova.

Once the nucleosynthesis process arrives at iron-56, the continuation of this process consumes energy (the addition of fragments to nuclei releases less energy than required to break them off the parent nuclei). If the mass of the core exceeds the Chandrasekhar limit, electron degeneracy pressure will be unable to support its weight against the force of gravity, and the core will undergo sudden, catastrophic collapse to form a neutron star or a black hole.

Through a process that is not completely understood, some of the gravitational potential energy released by this core collapse is converted into a Type Ib, Type Ic, or Type II supernova. It is known that the core collapse produces a massive surge of neutrinos, as observed with supernova SN 1987A.

The extremely energetic neutrinos fragment some nuclei; some of their energy is consumed in releasing nucleons, including neutrons, and some of their energy is transformed into heat and kinetic energy, thus augmenting the shock wave started by rebound of some of the infalling material from the collapse of the core.

Electron capture in very dense parts of the infalling matter may produce additional neutrons. As some of the rebounding matter is bombarded by the neutrons, some of its nuclei capture them, creating a spectrum of heavier-than-iron material including the radioactive elements up to (and likely beyond) uranium.

Although non-exploding red giant stars can produce significant quantities of elements heavier than iron using neutrons released in side reactions of earlier nuclear reactions, the abundance of elements heavier than iron (and in particular, of certain isotopes of elements that have multiple stable or long-lived isotopes) produced in such reactions is quite different from that produced in a supernova.

Neither abundance alone matches that found in our solar system, so both supernovae and ejection of elements from red giant stars are required to explain the observed abundance of heavy elements and isotopes thereof.

The energy transferred from collapse of the core to rebounding material not only generates heavy elements, but (by a mechanism which is not fully understood) provides for their acceleration well beyond escape velocity, thus causing a Type Ib, Type Ic, or Type II supernova.

Note that current understanding of this energy transfer is still not satisfactory; although current models of Type Ib, Type Ic, and Type II supernovae account for part of the energy transfer, they are not able to account for enough energy transfer to produce the observed ejection of material.

Some evidence gained from analysis of the mass and orbital parameters of binary neutron stars (which require two such supernovae) hints that the collapse of an oxygen-neon-magnesium core may produce a supernova that differs observably (in ways other than size) from a supernova produced by the collapse of an iron core.

The most massive stars may be completely destroyed by a supernova with an energy greatly exceeding its gravitational binding energy. This rare event, caused by pair-instability, leaves behind no black hole remnant.

2.4 STELLAR REMNANTS

There are three possible deaths for a star. The final stage for a star is determined by their masses. They are:

- White Dwarfs
- Neutron Stars
- Black Holes

2.4.1 White Dwarfs

For a star of 1 solar mass, the resulting white dwarf is of about 0.6 solar masses, compressed into approximately the volume of the Earth.

White dwarfs are stable because the inward pull of gravity is balanced by the degeneracy pressure of the star's electrons. (This is a consequence of the Pauli's Exclusion Principle.)

Electron degeneracy pressure provides a rather soft limit against further compression; therefore, for a given chemical composition, white dwarfs of higher mass have a smaller volume. With no fuel left to burn, the star radiates its remaining heat into space for billions of years.

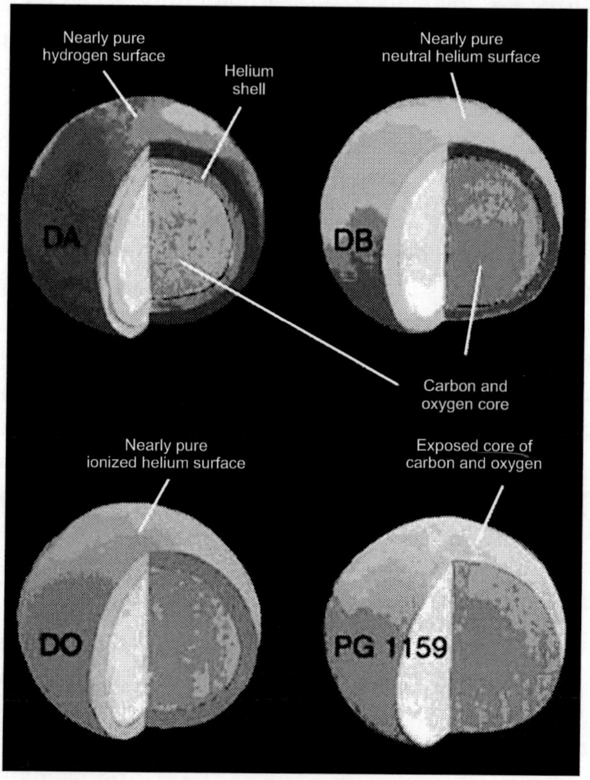

Fig. 2.2: Types of White Dwarfs

The chemical composition of the white dwarf depends upon its mass. A star of a few solar masses will ignite carbon fusion to form magnesium, neon, and smaller amounts of other elements, resulting in a white dwarf composed chiefly of oxygen, neon, and magnesium, provided that it can lose enough mass to get below the Chandrasekhar limit (*see* below), and provided that the ignition of carbon is not so violent as to blow apart the star in a supernova.

A star of mass on the order of magnitude of the Sun will be unable to ignite carbon fusion, and will produce a white dwarf composed chiefly of carbon and oxygen, and of mass too low to collapse unless matter is added to it later (see below). A star of less than about half the mass of the Sun will be unable to ignite helium fusion (as noted earlier), and will produce a white dwarf composed chiefly of helium.

In the end, all that remains is a cold dark mass sometimes called a black dwarf. However, the universe is not old enough for any black dwarf stars to exist yet. If the white dwarf's mass increases above the Chandrasekhar limit, which is 1.4 solar masses for a white dwarf, composed chiefly of carbon, oxygen, neon, and/or magnesium, then electron degeneracy pressure fails due to electron capture and the star collapses. Depending upon the chemical composition and pre-collapse temperature in the centre, this will either lead to collapse into a neutron star or runaway ignition of carbon and oxygen.

Heavier elements favour continued core collapse, because they require a higher temperature to ignite, because electron capture onto these elements and their fusion products is easier; higher core temperatures favour runaway nuclear reaction, which halts core collapse and leads to a Type Ia supernova. These supernovae may be many times brighter than the Type II supernova marking the death of a massive star, even though the latter has the greater total energy release. This instability to collapse means that no white dwarf more massive than approximately 1.4 solar masses can exist (with a possible minor exception for very rapidly spinning white dwarfs, whose centrifugal force due to rotation partially counteracts the weight of their matter). Mass transfer in a binary system may cause an initially stable white dwarf to surpass the Chandrasekhar limit.

If a white dwarf forms a close binary system with another star, hydrogen from the larger companion may accrete around and onto a white dwarf until it gets hot enough to fuse in a runaway reaction at its surface, although the white dwarf remains below the Chandrasekhar limit. Such an explosion is termed as nova.

2.4.2 Neutron Stars

When a stellar core collapses, the pressure causes electron capture, thus converting the great majority of the protons into neutrons. The electromagnetic forces keeping separate nuclei apart are gone (proportionally, if nuclei were the size of dust motes, atoms would be as large as football stadiums), and most of the core of the star becomes a dense ball of contiguous neutrons (in some ways like a giant atomic nucleus), with a thin overlying layer of degenerate matter (chiefly iron unless matter of different composition is added later). The neutrons resist further compression by the Pauli's Exclusion Principle, in a way analogous to electron degeneracy pressure, but stronger.

These stars, known as neutron stars, are extremely small — on the order of radius 10 km, no bigger than the size of a city — and are phenomenally dense. Their period of revolution shortens dramatically as the star shrinks (due to conservation of angular momentum); some spin at over 600 revolutions per second.

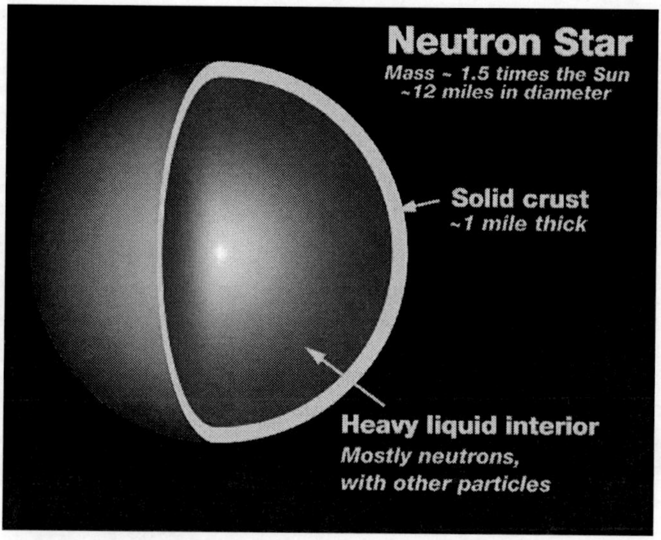

Fig. 2.3: Cross-section of a Neutron Star

The neutron stars would also be expected to have very high magnetic fields which could be trillion times the earth's field. In fact more than 30 years ago, pulsars were discovered emitting pulses of radio signals at very regular intervals. These proved to be the rapidly rotating neutron stars.

2.4.3 Black Holes

If the mass of the stellar remnant is high enough, the neutron degeneracy pressure will be insufficient to prevent collapse below the Schwarzschild radius. The stellar remnant thus becomes a black hole. The mass at which this occurs is not known with certainty, but is currently estimated at between 2 and 3 solar masses. Black holes are predicted by the theory of general relativity.

When Chandrasekhar presented his results (that the maximum degenerate stable mass of 1.4 solar mass) to Eddington, he was very critical. He told Chandrasekhar that his results "implied that a star will keep on shrinking till no light comes out of it, I think that something must intervene to prevent this absurd occurrence!"

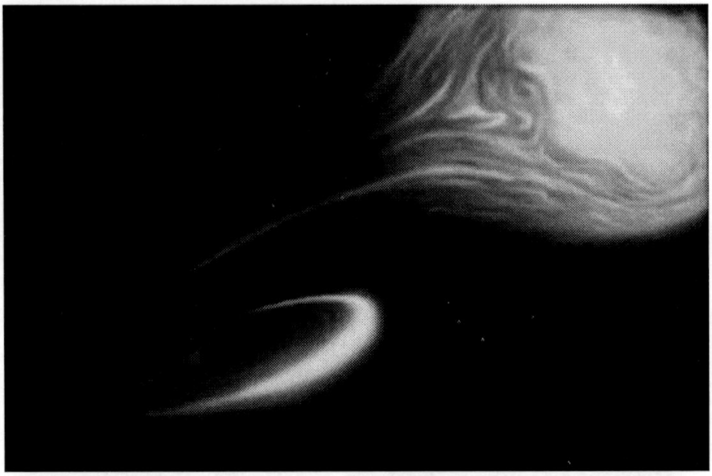

Fig 2.4: Accretion disc around a black hole

According to classical general relativity, no matter or information can flow from the interior of a black hole to an outside observer, although quantum effects may allow deviations from this strict rule. The existence of black holes in the universe is well-supported, both theoretically and by astronomical observation.

Since the core-collapse supernova mechanism itself is imperfectly understood, it is still not known whether it is possible for a star to collapse directly to a black hole without producing a visible supernova, or whether some supernovae initially form unstable neutron stars which then collapse into black holes; the exact relation between the initial mass of the star and the final remnant is also not completely certain. Resolution of these uncertainties requires the analysis of more supernovae and supernova remnants.

3 Black Holes

The universe of black holes is opening up for astronomical observations and discoveries over all wavelengths. This chapter gives a brief summary of this exciting current area of interest starting from the very concept of such objects. The various properties of black holes and the physics underlying them are explained and consequences explored. Also recent astronomical evidences for such objects are described.

The idea of a black hole actually goes back to Laplace and may be even earlier to Englishman Michel. Laplace argued that the largest objects in the universe must become invisible and would hence be dark.

He arrived at this conclusion in the following interesting manner.

Imagine an object of the same average density as the sun, that is about $1.5 g/cc$. He argued that if such an object were about 300 times larger than the sun (that is with the same density), the escape velocity for the object (given by) $v_{es} = \sqrt{\frac{2GM}{R}}$ would exceed that of light.

In terms of the density and radius of the object the escape velocity can be written as $v_{es}^2 = \frac{8\pi}{3} G\rho R^2$.

So the critical value of R (for a given density) at which the escape velocity becomes the speed of light is given by:

$$R_{crit} = \sqrt{\frac{3c^2}{8\pi G\rho}} \qquad \dots (1)$$

So given any object of density ρ, there is an upper limit to R, above which light cannot escape from the object.

In other words, this can be stated that given an object of mass M, it would have the smallest possible radius of

$$Rs = \frac{2GM}{c^2} \qquad \dots (2)$$

at which its escape velocity equals the light velocity so that no radiation can leave the star. When Einstein proposed his general theory of relativity, Schwarzschild solved the equation for spherical star and the solution implied that below a radius called the Schwarzschild radius, which coincidentally agree with equation (2), all light and other radiation is trapped inside the star, and it becomes what is called a black hole!

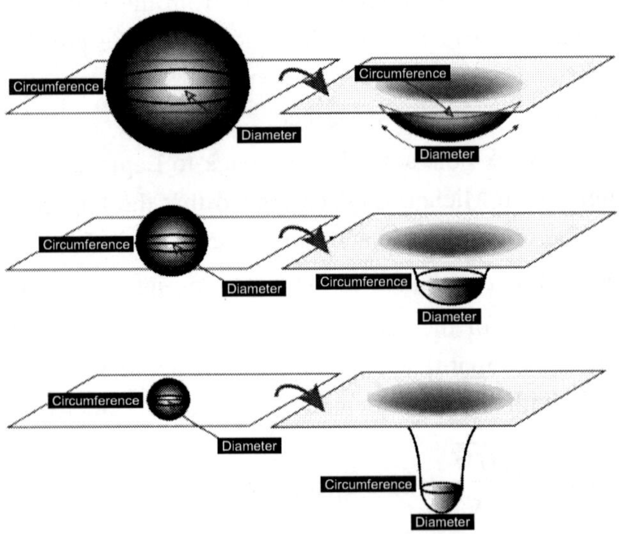

Fig. 3.1: Curvature of space-time around stars of same mass but different size. As the density of star increases, the space-time is bent to a larger extent.

In the case of a black hole, the gravitational field is so strong, that the ray of light are bent completely and keep going round in circles, so that the light and all other matter and radiation is completely trapped inside. This is different from the Newtonian picture where escape velocity was used. However the radius at which this trapping occurs is the same in both cases.

We know that a strong gravitational field bends light (in fact this was a major prediction of Einstein's general theory of relativity). For the sun this deflection for a ray of light just grazing its surface is about 1.75 seconds of an arc. This small angle of deflection was actually measured by Eddington's team at a total solar eclipse in 1919 and helped to make Einstein and his theory famous.

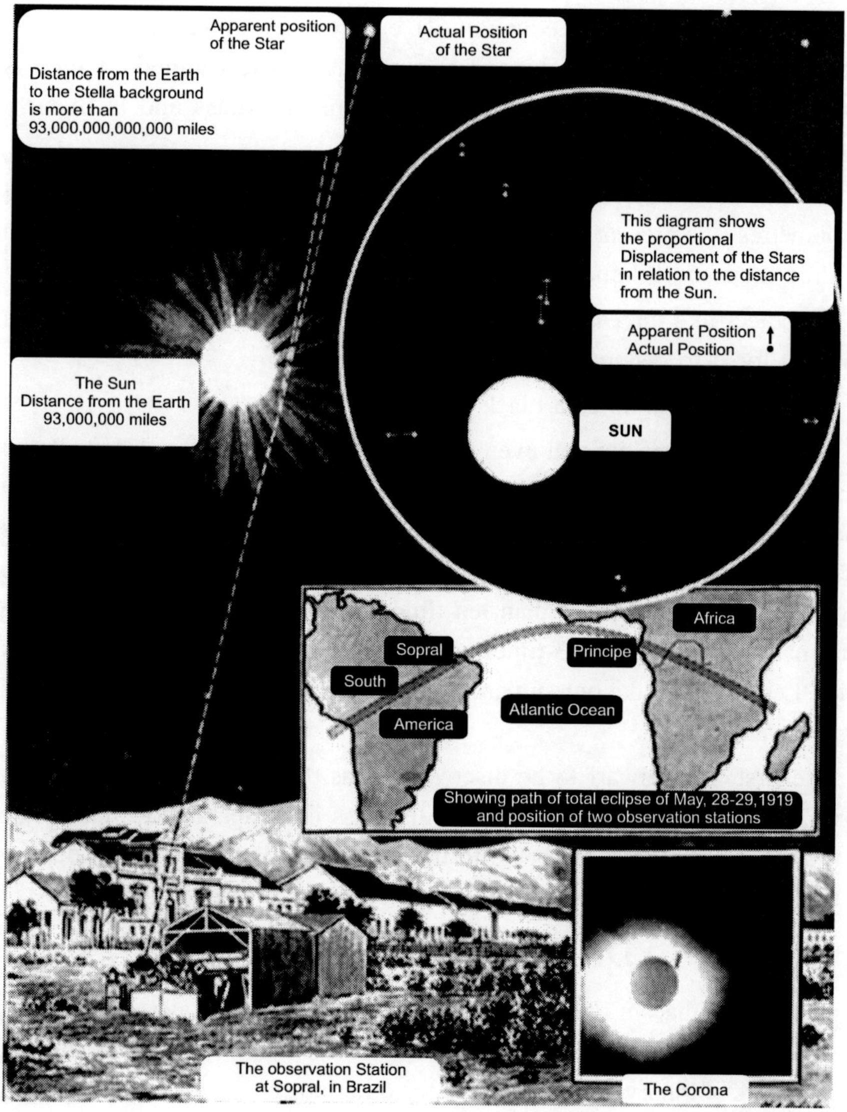

Fig.3.2: Famous experiment of Eddington during the total solar eclipse of 1919, which vindicated Einstein's general theory of relativity.

The term black hole became coined for popular universal usage apparently in an article by John Wheeler in Physics Today, 1971 "Introducing the Black Hole".

The formation of such a trapped surface as a star of mass M collapse beyond the radius given by equation (2), ensures that no matter or radiation can escape from the object once this radius of the trapped surface is reached. We see that this radius depends only on mass and is linearly proportional to it.

For a star of the sun's mass, this critical Schwarzschild radius is about 3 kilometres (the present radius at which it is shining steadily is about seven hundred thousand kilometres).

An object like the earth has to shrink to about one centimetre (size of a marble!) for it to become a black hole.

Will the sun end up as a black hole?

However, it turns out that even the sun is not massive enough to collapse finally into a black hole. Our present understanding of the evolution of stars (that is after they have exhausted their stock of nuclear fuel which powers them through thermonuclear reactions) is that only a massive star which to begin with is at least more than ten times heavier than the sun (and has a core mass of about 2.5 to 3 times solar mass) can collapse to form a black hole. 'Lesser' stars like our sun are destined to end up as what are known as white dwarfs.

The first white dwarf to be discovered was the companion star orbiting Sirius, but thousands of times fainter. Its mass is comparable to that of the sun, but has a radius comparable to the earth. This implies that the density of this star is few tons per cubic centimetre.

3.1 DETECTING BLACK HOLES

To be sure that an object is a black hole, we should have some means of measuring its mass. In our galaxy the several black hole candidates are all members of binary systems, where the other object is a usual star emitting visible radiation! This enables us to measure the mass of the invisible companion. In most cases, their masses turn out to be several solar masses, beyond that for neutron stars.

The black hole pulls out matter from its companion star; this matter forms an accretion disc around the hole and is slowly sucked in, the vast gravitational energy is converted into intense X-ray and other forms of electromagnetic radiation.

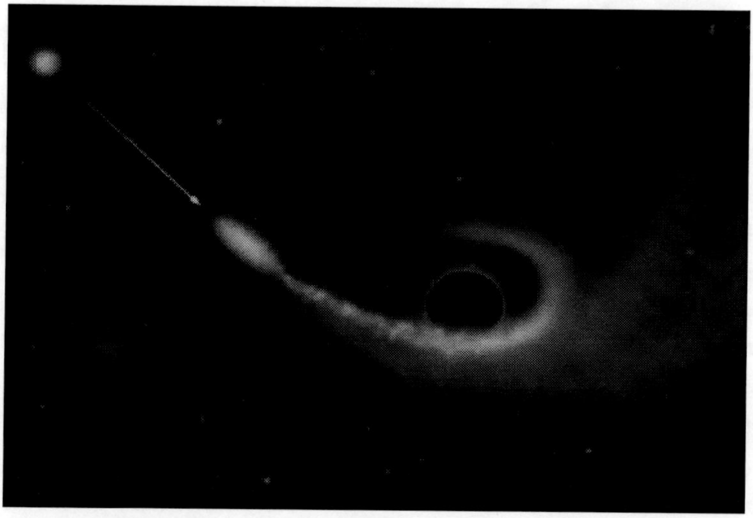

Fig. 3.3: Black hole accreting matter from a companion star (the star is ripped up due to high tidal force

More matter is pulled in when the star is closer in orbit around the hole, so that the high energy radiation (like X-rays) exhibits a periodicity corresponding to the binary period. This is the signature of black holes.

Neutron stars also accrete matter when part of binary system, from their companion stars and when this matter strikes the neutron star surface, X-rays are emitted. However, it is now possible to differentiate between a black hole and a neutron star. Matter when striking neutron star hits a 'hard' surface and can be emitted back with additional energy, whereas that falling into a black hole disappears into the horizon.

So the ratio of maximal to minimal luminosity is much higher for black holes, however massive the black hole for a given higher maximal luminosity. This relation between maximal luminosity and a given mass goes back to Eddington, who showed that given a mass of a luminous object, there is a maximum value for its luminosity as the radiation pressure would tend to push the matter apart exceeding the gravitational force supporting it. For a mass M in units of solar mass, this maximal (Eddington) luminosity is given by:

$$L_{Edd} = 10^{31} Watts \left(\frac{M}{M_{sun}} \right).$$

It is now believed that ultra luminous objects like quasars (which emit quadrillion times as much energy per second as the sun in all wavelengths!) are powered by supermassive black holes in their centre, accreting the surrounding matter, including gas, dust, stars, etc., releasing the gravitational energy as radiation.

The above formula shows that for a quasar to emit 10^{47}ergs/s, we need a supermassive black hole of about 10^9 solar mass, accreting matter at a rate of about one solar mass per year(!) to generate the required energy. Thermonuclear reactions which convert less than 0.7 per cent of matter to energy cannot account for the vast energy released by these objects.

There is also *direct* evidence for the presence of supermassive black holes in the centres of many galaxies including our own! We can measure the velocities of motions of the stars and gas in the innermost parsec of these galaxies, by Doppler broadening of their spectral emission lines as they go around the compact central object. From these velocities (typically several thousand km/s) we can estimate the total interior mass.

So one can deduce, for instance, that the inner 0.1 light years of our galaxy houses a mass of a few million solar masses. And only a black hole can neatly fit this bill! If we hypothesise that this region contains large number of massive objects this will pose several theoretical and observational problems.

Maser emissions from some spiral galaxies have enabled an accurate estimate of the rotational velocities of gas orbiting the massive compact central object. So now we have strong evidence for the existence of supermassive black holes, powering the most energetic objects in the universe like the quasars.

Galaxies with intense star forming regions (or star burst regions) like M82, show evidence for black holes of several hundred or thousand solar masses from their X-ray emission.

3.2 PRIMORDIAL BLACK HOLES

A primordial black hole is a hypothetical type of black hole that is formed not by the gravitational collapse of a star but by the extreme densities of matter present during early universe.

In the first few moments after the big bang, pressure and temperature were extremely great. Under these conditions, simple fluctuations in the density of matter may have resulted in local regions dense enough to create black holes. Although most regions of high density would be quickly dispersed by the expansion of the universe, a primordial black hole would be stable, persisting to the present.

One way to detect primordial black holes is by their Hawking radiation. Although classically an isolated black hole will not emit any light or radiation, Hawking showed that there is a quantum effect wherein radiation can tunnel out of the strong gravitational potential around the black hole. The gravitational field is strong enough to create particles. He also showed that the emission is thermal and associated a temperature with the black hole which is inversely proportional to its mass.

This is given by:

$$T_{BH} = 10^{-7} K \left(\frac{M}{M_{sun}} \right)$$

For solar mass and heavier black holes this radiation is extremely small, less than 10^{-17} ergs/s.

Since this emission further decreases their mass, black holes with very small mass would experience runaway evaporation, creating a massive burst of radiation.

If the black hole is losing mass then, its Schwarzschild radius must be decreasing. This equates to a decrease in the surface area of the event horizon. This would seem to violate Hawking's own area theorem, which states that the area can never decrease. The area theorem of general relativity gets replaced by a second law of thermodynamics for black holes, which states that the sum of the entropies of the black hole and the matter outside the black hole never decreases. While emitting a particle decreases the entropy of the black hole, the materialized particle has its own entropy that when summed together, equals or exceeds the initial entropy.

3.3 BLACK HOLES WITH SPIN AND CHARGE

Black holes could also have electric charge and be rotating. A black hole having an angular momentum J and a mass M has a radius given by:

$$r = m \pm \sqrt{m^2 - a^2}$$

Here $m = \frac{GM}{c^2}$ is the geometric mass and $a = \frac{J}{Mc}$ is the geometric angular momentum. From the condition that r should be real, $m \geq a$ .This ensures that the horizon always survives and the formation of a naked singularity (so called cosmic censorship conjecture of Penrose!) is averted. No matter what we do to the black hole, its 'interior' is always protected from 'outside observers'.

Similar thing holds good for a charged or charged-rotating black holes. In these cases the horizon radius, respectively, is given as:

$$r = m \pm \sqrt{m^2 - q^2} \quad \text{and} \quad r = m \pm \sqrt{m^2 - a^2 - q^2}$$

Where $q^2 = \frac{Ge^2}{C^4}$

So here again to avoid naked singularity we must have $m^2 \geq q^2$ and $m^2 \geq a^2 + q^2$. Black holes that satisfy the equalities, that is have the maximal value $(m = a,\ m^2 = q^2\ and\ m^2 = a^2 + q^2)$ are called extremal black holes.

Compared to stars black holes are 'simple' objects! They can have only three measurable attributes as far as an outside observer is concerned. They are:

1. Total mass or energy
2. Total electric charge
3. Total angular momentum

An outside observer loses all trace of the type and nature of matter that went inside a black hole.

Imagine that you are being drawn into a black hole. At first, you don't feel any gravitational forces at all, since you're in free fall. As you get closer and closer to the centre of the hole, though, you start to feel "tidal" gravitational forces. Imagine that your feet are closer to the centre than your head.

The gravitational pull gets stronger as you get closer to the centre of the hole, so your feet feel a stronger pull than your head does. As a result you feel "stretched." These tidal forces get more and more intense as you get closer to the centre, and eventually they will rip you apart.

In particular, nothing special happens at the moment when you cross the horizon (except for the fact that you have been ripped apart by the tidal forces!). Even after you've crossed the horizon, you can still see things on the outside: after all, the light from the things on the outside can still reach you.

No one on the outside can see you, of course, since the light from you can't escape past the horizon. The event horizon thus acts as a one way membrane. But an outside observer sees things differently. As you get closer and closer to the horizon, the observer sees you move more and more slowly. In fact, no matter how long the observer waits, he will never quite see you reach the horizon.

Suppose that the black hole formed from a collapsing star. As the material that is to form the black hole collapses, the observer sees it get smaller and smaller, approaching but never quite reaching its Schwarzschild radius. It does not really take an infinite amount of time for the black hole to form, and it does not really take an infinite amount of time for you to cross the horizon.

As you get closer and closer to the horizon, the light that you're emitting takes longer and longer to climb back out to reach the observer. In fact, the radiation you emit right, as you cross the horizon will hover right there at the horizon forever and never reach the observer. You've long since passed through the horizon, but the light signal telling the observer would not reach him for an infinitely long time.

3.4 CAN BLACK HOLES BE MADE IN THE LAB?

September 10, 2008 was a landmark date for high energy particle physics as the first high energy beam of multi-TeV protons whizzed around the 27 km tunnel of the LHC, heralding what is presently the world's most powerful accelerator. One TeV is a terra electron volt, or the energy gained by an electron in an electric potential corresponding to a trillion volts.

A lot of consternation especially among the general public was evinced as to possible disastrous consequence of such spectacular experiments. A favourite theme which has caught the imagination is the possibility of black hole production in such high energy collisions. It is feared that once such a black hole is produced it would quickly accrete all the surrounding matter including the whole earth!

Such notions are totally untenable and to explain this, a quantitative understanding of the physics involved in black hole formation and estimates of the energies involved etc. are required.

To begin with one of the reasons to be excited about the very high collision energies (~14TeV) of the protons in the oppositely moving beams is that it corresponds to the energies of the particle in the very early (high temperature, high density) phase of the universe, or to be more precise, one picosecond after the universe started expanding (in the big bang) the temperature corresponded to about a few TeV.

The LHC energies correspond to the particle energies about ten femtoseconds after the universe began expanding. In the early universe primordial black holes are formed when the gravitational potential equals the square of the velocity of light. This could happen, for example, if in the radiation dominated era, the external radiation pressure forced material inside the so called gravitational radius provided it began with a density sufficiently in excess of ambient average density.

It turns out that the mass of the black hole that can form in the early universe depends on the epoch when it formed. This shows that primordial black holes of around the mass of the earth could have formed in the universe when the temperatures (energies) were several TeV (that is at a time of one picosecond after the universe started expanding).

So does this mean that the LHC can produce earth mass black holes?

Let us consider the total energy required to form an earth mass black hole. We should remember that in the big bang when the temperature of the universe was a few TeV, the size of the universe (at that time) was hundred billion metres!

This entire volume (of $10^{33}m^3$) was at a temperature of $10^{17}K$! (That is, the volume of about the solar system was filled with quanta and particles of energy of many TeV, with a total energy content of ($\sim 10^{82}$J!). Whereas in the LHC, we have particles with individual energies (~TeV) just confined to the vacuum tubes of a highly localised 27 km region. To produce a black hole of the earth mass, we would have to squeeze a total energy of $\sim 10^{41}$J, in a volume of $10^{-5}m^3$, that is an earth mass black hole would have a horizon radius of just a centimetre!

With our world total power output of 10^{13}J/s, we would have to produce power at this rate for 10^{20}years (ten billion times the age of the universe!) to produce this much of energy and squeeze it into a region of one cubic centimetre! We need 10^{48}, TeV energy protons, to be squeezed in a region of a cubic centimetre. What we actually have in the LHC, is something thirty orders smaller! (We need a septillion grams of TeV energy protons!)

The pressure required is $\sim 10^{41}$ atmospheres! This is thirty orders ($\sim 10^{30}$ times) of magnitude more than what can be produced with the most powerful lasers in the world! And to produce smaller black holes, we need much higher pressures and temperatures! The energy density scales as $\frac{1}{M_{BH}^{2}}$.

At the worst, the high energy beams in the LHC can go out of control and damage only accelerator and surrounding structures! The total energy is just too low! The individual particle energies (temperatures) are high. (In a fluorescent lamp, the temperature of the individual particles is several thousand degrees, but the tube is cold to the touch, as the total heat content is low).

As we saw the energy required to form an earth mass black hole corresponds to our world's power production for years and all this has to be concentrated in a region of one cubic centimetre and to do that we have to squeeze it with a pressure of atmospheres.

The so called Hawking black holes of asteroid mass, were formed in the early universe when the temperatures were, eight orders higher than the particle energies in the LHC.

The energies required to form such black holes is again! Still smaller black holes require even higher particle energies and temperatures! So the formation of black holes in high energy particle collisions requires total energies and particle energies far beyond any contemporary technological endeavours!

The only black holes which could perhaps be produced (may be even copiously) are the TeV mass black holes (weighing). These could form, if there are additional space dimensions, and gravity and electroweak interactions unify at energies of several TeV.

However, such TeV mass black holes are likely to decay very fast on time scales ~yocto-second (or less) into a plethora of particle jets. So even if such black holes are produced, they would decay fast and not grow at all. No danger from such 'objects'!

In short an understanding of the basic physics involved in black hole formation shows that all fears of such objects forming and posing threats to mankind is totally unfounded. Again the Fermilab has been colliding TeV protons for some years now and cosmic ray collisions with much higher energies (up to) have been going on all over the universe for aeons (with no damage!)

□□□

4

Galaxies

A galaxy is a massive, gravitationally bound system consisting of stars, an interstellar medium of gas and dust, and dark matter. Typical galaxies range from dwarfs with as few as ten million (10^7) stars up to giants with one trillion (10^{12}) stars, all orbiting a common centre of mass. Galaxies can also contain many multiple star systems, star clusters, and various interstellar clouds. The Sun is one of the stars in the Milky Way galaxy; the Solar System includes the Earth and all the other objects that orbit the Sun.

Historically, galaxies have been categorised according to their apparent shape (usually referred to as their visual morphology). A common form is the elliptical galaxy, which has an ellipse-shaped light profile. Spiral galaxies are disk-shaped assemblages with curving, dusty arms. Galaxies with irregular or unusual shapes are known as peculiar galaxies, and typically result from disruption by the gravitational pull of neighbouring galaxies.

Such interactions between nearby galaxies, which may ultimately result in galaxies merging, may induce episodes of significantly increased star formation, producing what is called a starburst galaxy. Small galaxies that lack a coherent structure could also be referred to as irregular galaxies.

There are probably more than 100 billion (10^{11}) galaxies in the observable universe. Most galaxies are 1,000 to 100,000 parsecs in diameter and are usually separated by distances on the order of millions of parsecs (or mega parsecs). Intergalactic space (the space between galaxies) is filled with a tenuous gas of an average density less than one atom per cubic meter.

The majority of galaxies are organised into a hierarchy of associations called clusters, which, in turn, can form larger groups called superclusters. These larger structures are generally arranged into sheets and filaments, which surround immense voids in the universe.

Although it is not yet well-understood, dark matter appears to account for around 90% of the mass of most galaxies. Observational data suggests that supermassive black holes may exist at the centre of many, if not all, galaxies.

They are proposed to be the primary cause of active galactic nuclei found at the core of some galaxies. The Milky Way galaxy appears to harbour at least one such objects within its nucleus.

4.1 TYPES OF GALAXIES

Galaxies come in three main types: ellipticals, spirals, and irregulars. A slightly more extensive description of galaxy types based on their appearance is given by the Hubble sequence. Since the Hubble sequence is entirely based upon visual morphological type, it may miss certain important characteristics of galaxies such as star formation rate (in starburst galaxies) and activity in the core (in active galaxies).

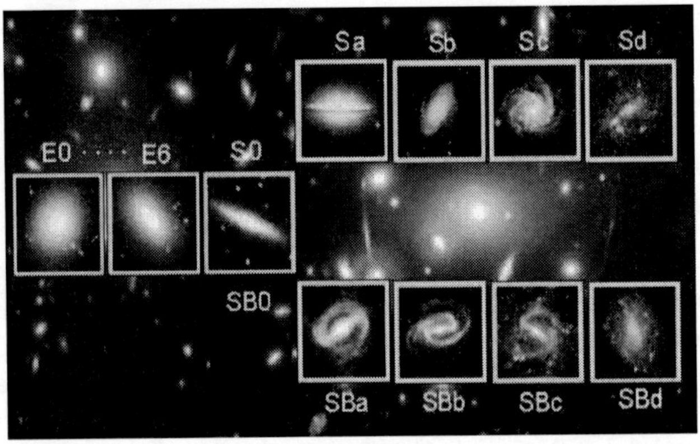

Fig. 4.1: Types of galaxies

4.1.1 Elliptical Galaxies

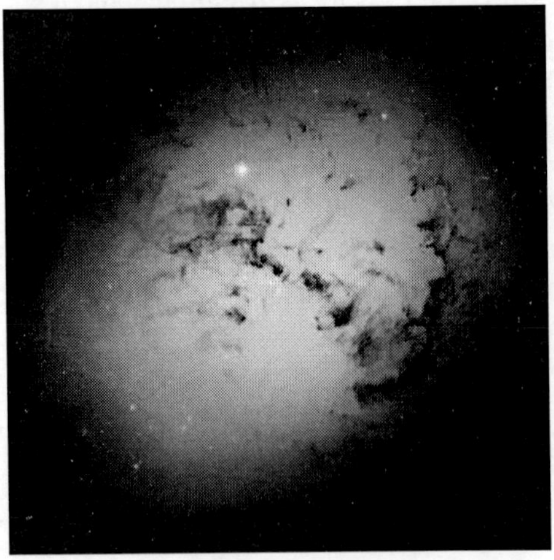

Fig. 4.2: Elliptical galaxy

The Hubble classification system rates elliptical galaxies on the basis of their ellipticity, ranging from E0, being nearly spherical, up to E7, which is highly elongated. These galaxies have an ellipsoidal profile, giving them an elliptical appearance regardless of the viewing angle. Their appearance shows little structure and they typically have relatively little interstellar matter.

Consequently these galaxies also have a low portion of open clusters and a reduced rate of new star formation. Instead the galaxy is dominated by generally older, more evolved stars that are orbiting the common centre of gravity in random directions. In this sense they have some similarity to the much smaller globular clusters.

The largest galaxies are giant ellipticals. Many elliptical galaxies are believed to form due to the interaction of galaxies, resulting in a collision and merger. They can grow to enormous sizes (compared to spiral galaxies, for example), and giant elliptical galaxies are often found near the core of large galaxy clusters. Starburst galaxies are the result of such a galactic collision that can result in the formation of an elliptical galaxy.

4.1.2 Spiral Galaxies

Spiral galaxies consist of a rotating disk of stars and interstellar medium, along with a central bulge of generally older stars. Extending outward from the bulge are relatively bright arms. In the Hubble classification scheme, spiral galaxies are listed as type *S*, followed by a letter (*a*, *b*, or *c*) that indicates the degree of tightness of the spiral arms and the size of the central bulge. An *Sa* galaxy has tightly wound poorly-defined arms and possesses a relatively large core region. A *Sc* galaxy has open, well-defined arms and a small core region.

Fig.4.3: Spiral galaxy

In spiral galaxies, the spiral arms do have the shape of approximate logarithmic spirals, a pattern that can be theoretically shown to result from a disturbance in a uniformly rotating mass of stars. Like the stars, the spiral arms also rotate around the centre, but they do so with constant angular velocity. That means that stars pass in and out of spiral arms, with stars near the galactic core orbiting faster than the arms are moving while stars near the outer parts of the galaxy typically orbit more slowly than the arms.

The spiral arms are thought to be areas of high density matter, or "density waves". As stars move through an arm, the space velocity of each stellar system is modified by the gravitational force of the higher density. (The velocity returns to normal after the stars depart on the other side of the arm.) This effect is akin to a "wave" of slowdowns moving along a highway full of moving cars. The arms are visible because the high density facilitates star formation, and therefore they harbour many bright and young stars.

A majority of spiral galaxies have a linear, bar-shaped band of stars that extends outward to either side of the core, then merges into the spiral arm structure. In the Hubble classification scheme, these are designated by an *SB*, followed by a lower-case letter (*a*, *b* or *c*) that indicates the form of the spiral arms (in the same manner as the categorisation of normal spiral galaxies).

Bars are thought to be temporary structures that can occur as a result of a density wave radiating outward from the core, or else due to a tidal interaction with another galaxy. Many barred spiral galaxies are active, possibly as a result of gas being channelled into the core along the arms.

Our own galaxy is a large disk-shaped barred-spiral galaxy about 30 kilo parsecs in diameter and a kilo parsec in thickness. It contains about two hundred billion (2×10^{11}) stars and has a total mass of about six hundred billion (6×10^{11}) times the mass of the Sun.

4.1.3 Dwarf Galaxies

Despite the prominence of large elliptical and spiral galaxies, most galaxies in the universe appear to be dwarf galaxies. These tiny galaxies are about one hundredth the size of the Milky Way, containing only a few billion stars. Ultra-compact dwarf galaxies have recently been discovered that are only 100 parsecs across. Many dwarf galaxies may orbit a single larger galaxy; the Milky Way has at least a dozen such satellites, with an estimated 300–500 yet to be discovered.

Dwarf galaxies may also be classified as elliptical, spiral, or irregular. Since small dwarf ellipticals bear little resemblance to large ellipticals, they are often called dwarf spheroidal galaxies instead. A study of 27 Milky Way neighbours found that dwarf galaxies were all approximately 10 million solar masses, regardless of whether they have thousands or millions of stars. This has led to the suggestion that galaxies are largely formed by dark matter, and that the minimum size may indicate a form of warm dark matter incapable of gravitational coalescence on a smaller scale.

4.1.4 Starburst Galaxies

Stars are created within galaxies from a reserve of cold gas that forms into giant molecular clouds. Some galaxies have been observed to form stars at an exceptional rate, known as a starburst. Should they continue to do so, however, they would consume their reserve of gas in a time frame lower than the lifespan of the galaxy. Hence starburst activity usually lasts for only about ten million years, a relatively brief period in the history of a galaxy. Starburst galaxies were more common during the early history of the universe, and, at present, still contribute an estimated 15% to the total star production rate.

Starburst galaxies are characterised by dusty concentrations of gas and the appearance of newly-formed stars, including massive stars that ionize the surrounding clouds to create H II regions. These massive stars also produce supernova explosions, resulting in expanding remnants that interact powerfully with the surrounding gas. These outbursts trigger a chain reaction of star building that spreads throughout the gaseous region. Only when the available gas is nearly consumed or dispersed does the starburst activity come to an end.

Fig. 4.4: Starburst galaxy

Starbursts are often associated with merging or interacting galaxies. The prototype example of such a starburst-forming interaction is M82, which experienced a close encounter with the larger M81. Irregular galaxies often exhibit spaced knots of starburst activity.

4.2 ACTIVE GALACTIC NUCLEUS

A portion of the galaxies we can observe are classified as active. That is, a significant portion of the total energy output from the galaxy is emitted by a source other than the stars, dust and interstellar medium. The standard model for an active galactic nucleus is based upon an accretion disc that forms around a supermassive black hole (SMBH) at the core region.

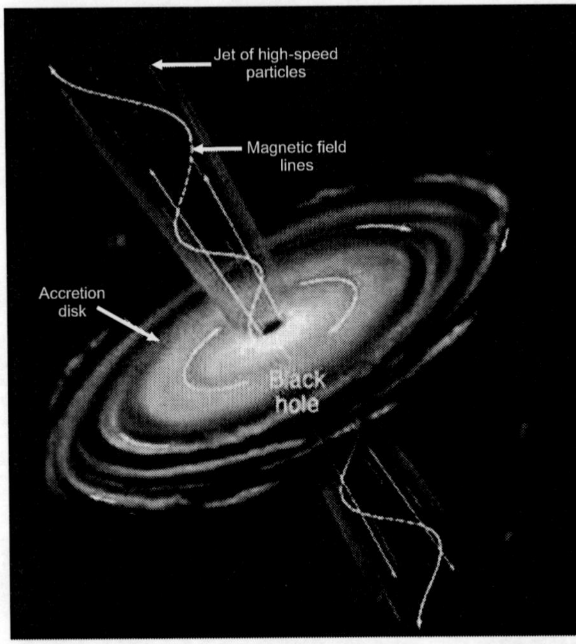

Fig. 4.5: AGN with the central black hole and the high energy jets

The radiation from an active galactic nucleus results from the gravitational energy of matter as it falls toward the black hole from the disc. In about 10% of these objects, a diametrically opposed pair of energetic jets ejects particles from the core at velocities close to the speed of light. The mechanism for producing these jets is still not well-understood.

Active galaxies that emit high-energy radiation in the form of x-rays are classified as Seyfert galaxies or quasars, depending on the luminosity. Blazars are believed to be an active galaxy with a relativistic jet that is pointed in the direction of the Earth. A radio galaxy emits radio frequencies from relativistic jets. A unified model of these types of active galaxies explains their differences based on the viewing angle of the observer.

4.3 FORMATION AND EVOLUTION OF GALAXIES

The study of galactic formation and evolution attempts to answer questions regarding how galaxies formed and their evolutionary path over the history of the universe. Some theories in this field have now become widely accepted, but it is still an active area in astrophysics.

4.3.1 Formation

Current cosmological models of the early Universe are based on the Big Bang theory. About 300,000 years after this event, atoms of hydrogen and helium began to form, in an event called recombination. Nearly all the hydrogen was neutral (non-ionized) and readily absorbed light, and no stars had yet formed. As a result this period has been called the "Dark Ages".

It was from density fluctuations (or anisotropic irregularities) in this primordial matter that larger structures began to appear. As a result, masses of baryonic matter started to condense within cold dark matter halos. These primordial structures would eventually become the galaxies we see today.

Evidence for the early appearance of galaxies was found in 2006, when it was discovered that the galaxy IOK-1 has an unusually high redshift of 6.96, corresponding to just 750 million years after the Big Bang and making it the most distant and primordial galaxy yet seen. While some scientists have claimed other objects (such as Abell 1835 IR1916) have higher redshifts (and therefore are seen in an earlier stage of the Universe's evolution), IOK-1's age and composition have been more reliably established. The existence of such early protogalaxies suggests that they must have grown in the so-called "Dark Ages".

The detailed process by which such early galaxy formation occurred is a major open question in astronomy. Theories could be divided into two categories: top-down and bottom-up. In top-down theories, proto-galaxies form in a large-scale simultaneous collapse lasting about one hundred million years. In bottom-up theories, small structures such as globular clusters form first, and then a number of such bodies accrete to form a larger galaxy. Modern theories must be modified to account for the probable presence of large dark matter halos.

Once protogalaxies began to form and contract, the first halo stars (called Population III stars) appeared within them. These were composed almost entirely of hydrogen and helium, and may have been massive. If so, these huge stars would have quickly consumed their supply of fuel and became supernovae, releasing heavy elements into the interstellar medium. This first generation of stars re-ionized the surrounding neutral hydrogen, creating expanding bubbles of space through which light could readily travel.

4.3.2 Evolution

Within a billion years of a galaxy's formation, key structures begin to appear. Globular clusters, the central supermassive black hole, and a galactic bulge of metal-poor Population II stars form. The creation of a supermassive black hole appears to play a key role in actively regulating the growth of galaxies by limiting the total amount of additional matter added. During this early epoch, galaxies undergo a major burst of star formation.

During the following two billion years, the accumulated matter settles into a galactic disc. A galaxy will continue to absorb infalling material from high velocity clouds and dwarf galaxies throughout its life. This matter is mostly hydrogen and helium. The cycle of stellar birth and death slowly increases the abundance of heavy elements, eventually allowing the formation of planets.

The evolution of galaxies can be significantly affected by interactions and collisions. Mergers of galaxies were common during the early epoch, and the majority of galaxies were peculiar in morphology. Given the distances between the stars, the great majority of stellar systems in colliding galaxies will be unaffected. However, gravitational stripping of the interstellar gas and dust that makes up the spiral arms produces a long train of stars known as tidal tails. Examples of these formations can be seen in NGC 4676 or the Antennae Galaxies.

As an example of such an interaction, the Milky Way galaxy and the nearby Andromeda Galaxy are moving toward each other at about 130 km/s, and—depending upon the lateral movements—the two may collide in about five to six billion years. Although the Milky Way has never collided with a galaxy as large as Andromeda before, evidence of past collisions of the Milky Way with smaller dwarf galaxies is increasing.

Such large-scale interactions are rare. As time passes, mergers of two systems of equal size become less common. Most bright galaxies have remained fundamentally unchanged for the last few billion years, and the net rate of star formation probably also peaked approximately ten billion years ago.

At present, most star formation occurs in smaller galaxies where cool gas is not so depleted. Spiral galaxies, like the Milky Way, only produce new generations of stars as long as they have dense molecular clouds of interstellar hydrogen in their spiral arms. Elliptical galaxies are already largely devoid of this gas, and so form no new stars. The supply of star-forming material is finite; once stars have converted the available supply of hydrogen into heavier elements, new star formation will come to an end.

The current era of star formation is expected to continue for up to one hundred billion years, and then the "stellar age" will wind down after about ten trillion to one hundred trillion years (10^{13}–10^{14} years), as the smallest, longest-lived stars in our astrosphere, tiny red dwarfs, begin to fade. At the end of the stellar age, galaxies will be composed of compact objects: brown dwarfs, white dwarfs that are cooling or cold ("black dwarfs"), neutron stars, and black holes. Eventually, as a result of gravitational relaxation, all stars will either fall into central supermassive black holes or be flung into intergalactic space as a result of collisions.

4.4 LARGER SCALE STRUCTURES

Deep sky surveys show that galaxies are often found in relatively close association with other galaxies. Solitary galaxies that have not significantly interacted with another galaxy of comparable mass during the past billion years are relatively scarce. Only about 5% of the galaxies surveyed have been found to be truly isolated; however, these isolated formations may have interacted and even merged with other galaxies in the past, and may still be orbited by smaller, satellite galaxies. Isolated galaxies can produce stars at a higher rate than normal, as their gas is not being stripped by other, nearby galaxies.

On the largest scale, the universe is continually expanding, resulting in an average increase in the separation between individual galaxies (*see* Hubble's law). Associations of galaxies can overcome this expansion on a local scale through their mutual gravitational attraction. These associations formed early in the universe, as clumps of dark matter pulled their respective galaxies together. Nearby groups later merged to form larger-scale clusters. This ongoing merger process (as well as an influx of infalling gas) heats the intergalactic gas within a cluster to very high temperatures, reaching 30–100 million K. About 70–80% of the mass in a cluster is in the form of dark matter, with 10–30% consisting of this heated gas and the remaining few per cent of the matter in the form of galaxies.

□□□

5 Dark Matter and Dark Energy

The visible (baryonic) matter in the universe makes up only about 4% of all the energy density of the universe. Of the remaining, dark matter contributes about 23% and dark energy dominates with 73%. This chapter gives a brief account of dark matter as well as dark energy and possible candidates for both.

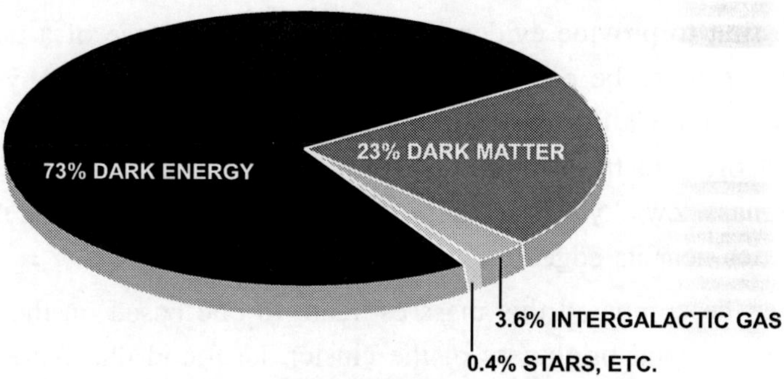

Fig. 5.1: Energy density of the universe

5.1 DARK MATTER

Dark matter is matter that does not interact with the electromagnetic force, but whose presence can be inferred from gravitational effects on visible matter. According to present observations of structures larger than galaxies, as well as Big Bang cosmology, dark matter and dark energy account for the vast majority of the mass in the observable universe.

The observed phenomena which imply the presence of dark matter include the rotational speeds of galaxies, orbital velocities of galaxies in clusters, gravitational lensing of background objects by galaxy clusters such as the Bullet cluster, and the temperature distribution of hot gas in galaxies and clusters of galaxies.

Dark matter also plays a central role in structure formation and galaxy evolution, and has measurable effects on the anisotropy of the cosmic microwave background. All these lines of evidence suggest that galaxies, clusters of galaxies, and the universe as a whole contain far more matter than that which interacts with electromagnetic radiation: the remainder is called the "dark matter component".

The dark matter component has much more mass than the "visible" component of the universe. At present, the density of ordinary baryons and radiation in the universe is estimated to be equivalent to about one hydrogen atom per cubic meter of space. Only about 4% of the total energy density in the universe (as inferred from gravitational effects) can be seen directly.

About 22% is thought to be composed of dark matter. The remaining 74% is thought to consist of dark energy, an even stranger component, distributed diffusely in space. Some hard-to-detect baryonic matter is believed to make a contribution to dark matter but would constitute only a small portion.

The first to provide evidence and infer the existence of a phenomenon that has come to be called "dark matter" was Swiss astrophysicist Fritz Zwicky, of the California Institute of Technology in 1933. He applied the virial theorem to the Coma cluster of galaxies and obtained evidence of unseen mass. Zwicky estimated the cluster's total mass based on the motions of galaxies near its edge.

When he compared this mass estimate to one based on the number of galaxies and total brightness of the cluster, he found that there was about 400 times more mass than expected. The gravity of the visible galaxies in the cluster would be far too small for such fast orbits, so something extra was required. This is known as the "missing mass problem". Based on these conclusions, Zwicky inferred that there must be some non-visible form of matter which would provide enough of the mass and gravity to hold the cluster together.

Much of the evidence for dark matter comes from the study of the motions of galaxies. Many of these appear to be fairly uniform, so by the virial theorem the total kinetic energy should be half the total gravitational binding energy of the galaxies. Experimentally, however, the total kinetic energy is found to be much greater: in particular, assuming the gravitational mass is due to only the visible matter of the galaxy; stars far from the centre of galaxies have much higher velocities than predicted by the virial theorem.

Galactic rotation curves, which illustrate the velocity of rotation versus the distance from the galactic centre, cannot be explained by only the visible matter. Assuming that the visible material makes *up* only a small part of the cluster is the most straightforward way of accounting for this. Galaxies show signs of being composed largely of a roughly spherically symmetric, centrally concentrated halo of dark matter with the visible matter concentrated in a disc at the centre.

Low surface brightness dwarf galaxies are important sources of information for studying dark matter, as they have an uncommonly low ratio of visible matter to dark matter, and have few bright stars at the centre which impair observations of the rotation curve of outlying stars.

For 40 years after Zwicky's initial observations, no other corroborating observations indicated that the mass to light ratio was anything other than unity (a high mass-to-light ratio indicates the presence of dark matter). Then, in the late 1960s and early 1970s, Vera Rubin, a young astronomer at the Department of Terrestrial Magnetism at the Carnegie Institution of Washington presented findings based on a new sensitive spectrograph that could measure the velocity curve of edge-on spiral galaxies to a greater degree of accuracy than had ever before been achieved.

Together with fellow staff-member Kent Ford, Rubin announced at a 1975 meeting of the American Astronomical Society the astonishing discovery that most stars in spiral galaxies orbit at roughly the same speed, which implied that their mass densities were uniform well beyond the locations with most of the stars (the galactic bulge). This result suggests that either Newtonian gravity does not apply universally or that, conservatively, upwards of 50% of the mass of galaxies was contained in the relatively dark galactic halo.

Though met with scepticism, Rubin insisted that the observations were correct. Eventually other astronomers began to corroborate her work and it soon became well-established that most galaxies were in fact dominated by "dark matter"; exceptions appeared to be galaxies with mass-to-light ratios close to that of stars. Subsequent to this, numerous observations have been made that do indicate the presence of dark matter in various parts of the cosmos.

Together with Rubin's findings for spiral galaxies and Zwicky's work on galaxy clusters, the observational evidence for dark matter has been collecting over the decades to the point that today most astrophysicists accept its existence. As a unifying concept, dark matter is one of the dominant features considered in the analysis of structures on the order of galactic scale and larger.

5.1.1 Velocity Dispersions of Galaxies

Rubin's pioneering work has stood the test of time. Measurements of velocity curves in spiral galaxies were soon followed up with velocity dispersions of elliptical galaxies. While sometimes appearing with lower mass-to-light ratios, measurements of ellipticals still indicate a relatively high dark matter content.

Likewise, measurements of the diffuse interstellar gas found at the edge of galaxies indicate not only dark matter distributions that extend beyond the visible limit of the galaxies, but also that the galaxies are virialized up to ten times their visible radii. This has the effect of pushing up the dark matter as a fraction of the total amount of gravitating matter from 50% measured by Rubin to the now accepted value of nearly 95%.

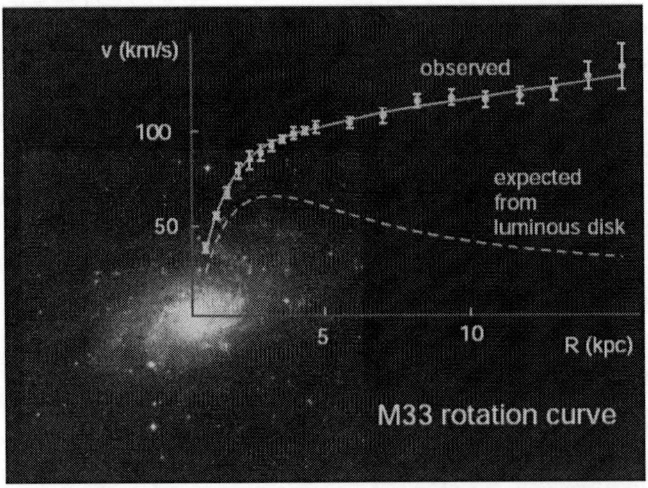

Fig. 5.2: Rotational curve for galaxy M33

There are places where dark matter seems to be a small component or totally absent. Globular clusters show no evidence that they contain dark matter, though their orbital interactions with galaxies do show evidence for galactic dark matter. For some time, measurements of the velocity profile of stars seemed to indicate concentration of dark matter in the disk of the

Milky Way galaxy, however, now it seems that the high concentration of baryonic matter in the disk of the galaxy (especially in the interstellar medium) can account for this motion. Galaxy mass profiles are thought to look very different from the light profiles.

The typical model for dark matter galaxies is a smooth, spherical distribution in virialized halos. Such would have to be the case to avoid small-scale (stellar) dynamical effects. Recent research reported in January, 2006 from the University of Massachusetts, Amherst would explain the previously mysterious warp in the disk of the Milky Way by the interaction of the Large and Small Magellanic Clouds and the predicted 20 fold increase in mass of the Milky Way taking into account dark matter.

In 2005, astronomers from Cardiff University claimed to discover a galaxy made almost entirely of dark matter, 50 million light years away in the Virgo Cluster, which was named VIRGOHI21. Unusually, VIRGOHI21 does not appear to contain any visible stars: it was seen with radio frequency observations of hydrogen.

Based on rotation profiles, the scientists estimate that this object contains approximately 1000 times more dark matter than hydrogen and has a total mass of about 1/10th that of the Milky Way Galaxy we live in.

For comparison, the Milky Way is believed to have roughly 10 times as much dark matter as ordinary matter. Models of the Big Bang and structure formation have suggested that such dark galaxies should be very common in the universe, but none had previously been detected.

If the existence of this dark galaxy is confirmed, it provides strong evidence for the theory of galaxy formation and poses problems for alternative explanations of dark matter.

Recently too there is evidence that there are 10 to 100 times fewer small galaxies than permitted by what the dark matter theory of galaxy formation predicts. There are also a small number of galaxies, like NGC 3379 whose measured orbital velocity of its gas clouds, show that it contains almost no dark matter at all.

Non-baryonic dark matter is divided into three different types:

- Hot dark matter - non-baryonic particles that move ultrarelativistically
- Warm dark matter - non-baryonic particles that move relativistically
- Cold dark matter - non-baryonic particles that move non-relativistically

5.1.2 Detection of Dark Matter

These cosmological models predict that if WIMPs are what make up dark matter, trillions must pass through the Earth each second. Despite a number of attempts to find these WIMPs, none have yet been confirmedly found.

Experimental searches for these dark matter candidates have been conducted and are ongoing. These efforts can be divided into two broad classes: direct detection, in which the dark matter particles are observed in a detector; and indirect detection, which looks for the products of dark matter annihilations.

Dark matter detection experiments have ruled out some WIMP and axion models. There are also several experiments claiming positive evidence for dark matter detection, such as DAMA/NaI, DAMA/LIBRA and EGRET, but these are so far unconfirmed and difficult to reconcile with the negative results of other experiments. Several searches for dark matter are currently underway, including the Cryogenic Dark Matter Search in the Soudan mine, the XENON, DAMA/LIBRA and CRESST experiments at Gran Sasso and the ZEPLIN and DRIFT projects at the Boulby Underground Laboratory (UK), and many new technologies are under development, such as the ArDM experiment.

One possible alternative approach to the detection of WIMPs in nature is to produce them in the laboratory. Experiments with the Large Hadron Collider near Geneva may be able to detect the WIMPs. Because a WIMP only has negligible interactions with matter, it can be detected as missing energy and momentum. It is also possible that dark matter consists of very heavy hidden sector particles which only interact with ordinary matter *via* gravity.

The Cryogenic Dark Matter Search, in the Soudan Mine in Minnesota aims to detect the heat generated when ultracold germanium and silicon crystals are struck by a WIMP. The Gran Sasso National Laboratory at L'Aquila, in Italy, uses xenon to measure the flash of light that occurs on those rare occasions when a WIMP strikes a xenon nucleus. The results from April, 2007, using 15 kg of liquid and gaseous xenon, detected several events consistent with backgrounds, setting a new exclusion limit. The larger XENON100 detector, with 150 kg of liquid xenon, began taking calibration data in March, 2008.

The Fermi space telescope, launched June 11, 2008, searching gamma wave events, may also detect WIMPs. WIMP supersymmetric particle and antiparticle collisions should release a pair of detectable gamma waves. The number of events detected will show to what extent WIMPs comprise dark matter.

With all these experiments together, scientists are becoming confident that WIMPs will be discovered in the near future. But some scientists are beginning to think that dark matter is composed of many different candidates. WIMPs may thus only be a part of the solution.

5.2 DARK ENERGY

Dark energy is a hypothetical exotic form of energy that permeates all of space and tends to increase the rate of expansion of the universe. Dark energy is the most popular way to explain recent observations that the universe appears to be expanding at an accelerating rate. In the standard model of cosmology, dark energy currently accounts for 74% of the total mass-energy of the universe.

Two proposed forms for dark energy are the cosmological constant, a *constant* energy density filling space homogeneously, and scalar fields such as quintessence or moduli, *dynamic* quantities whose energy density can vary in time and space. Contributions from scalar fields that are constant in space are usually also included in the cosmological constant. The cosmological constant is physically equivalent to vacuum energy. Scalar fields which do change in space can be difficult to distinguish from a cosmological constant because the change may be extremely slow.

High-precision measurements of the expansion of the universe are required to understand how the expansion rate changes over time. In general relativity, the evolution of the expansion rate is parameterized by the cosmological equation of state. Measuring the equation of state of dark energy is one of the biggest efforts in observational cosmology today.

Adding the cosmological constant to cosmology's standard metric leads to the Lambda-CDM model, which has been referred to as the "standard model" of cosmology because of its precise agreement with observations.

The existence of dark energy, in whatever form, is needed to reconcile the measured geometry of space with the total amount of matter in the universe. Measurements of cosmic microwave background (CMB) anisotropies, most recently by the WMAP satellite, indicate that the universe is very close to flat. For the shape of the universe to be flat, the mass/energy density of the universe must be equal to a certain critical density.

The total amount of matter in the universe (including baryons and dark matter), as measured by the CMB, accounts for only about 30% of the critical density. This implies the existence of an additional form of energy to account for the remaining 70%. The most recent WMAP observations are consistent with a universe made up of 73% dark energy, 23% dark matter, and 4% ordinary matter.

5.2.1 Nature of Dark Energy

The exact nature of this dark energy is a matter of speculation. It is known to be very homogeneous, not very dense and is not known to interact through any of the fundamental forces other than gravity.

Since it is not very dense—roughly 10^{-29} grams per cubic centimetre—it is hard to imagine experiments to detect it in the laboratory.

Dark energy can only have such a profound impact on the universe, making up 74% of all energy, because it uniformly fills otherwise empty space. The two leading models are quintessence and the cosmological constant. Both models include the common characteristic that dark energy must have negative pressure.

Independently from its actual nature, dark energy would need to have a strong negative pressure in order to explain the observed acceleration in the expansion rate of the universe.

According to General Relativity, the pressure within a substance contributes to its gravitational attraction for other things just as its mass density does. This happens because the physical quantity that causes matter to generate gravitational effects is the Stress-energy tensor, which contains both the energy (or matter) density of a substance and its pressure and viscosity.

In the Friedmann-Lemaître-Robertson-Walker metric, it can be shown that a strong constant negative pressure in the entire universe causes an acceleration in universe expansion if the universe is already expanding or a deceleration in universe contraction if the universe is already contracting.

More exactly, the second derivative of the universe scale factor, is positive if the equation of state of the universe is such that $w < 1 / 3$.

This accelerating expansion effect is sometimes labelled "gravitational repulsion", which is a colourful but possibly confusing expression. In fact a negative pressure does not influence the gravitational interaction between masses - which remains attractive - but rather alters the overall evolution of the universe at the cosmological scale, typically resulting in the accelerating expansion of the universe despite the attraction among the masses present in the universe.

5.2.2 Cosmolgical Constant

The simplest explanation for dark energy is that it is simply the "cost of having space": that is, a volume of space has some intrinsic, fundamental energy. This is the cosmological constant, sometimes called Lambda (hence Lambda-CDM model) after the Greek letter Ë, the symbol used to mathematically represent this quantity. Since energy and mass are related by $E = mc^2$, Einstein's theory of general relativity predicts that it will have a gravitational effect.

It is sometimes called a vacuum energy because it is the energy density of empty vacuum. In fact, most theories of particle physics predict vacuum fluctuations that would give the vacuum this sort of energy. This is related to the Casimir Effect, in which there is a small suction into regions where virtual particles are geometrically inhibited from forming (*e.g.*, between plates with tiny separation). The cosmological constant is estimated by cosmologists to be on the order of 10^{-29} g/cm^3, or about 10^{120} in reduced Planck units. However, particle physics predicts a natural value of 1 in reduced Planck units, a large discrepancy which is still lacking in explanation.

□□□

Astronomy Quiz

QUESTIONS

1. Which object in the solar system has the fastest frequency of rotation? What is its oblateness?
2. What is special about the star R23?
3. Asteroid No.7189 Kuniko is named after which person? He is known for what discovery?
4. A Japanese barber discovered a nova during the total solar eclipse of June 19, 1936. Name him and the nova he discovered.
5. Which French astronomer, who was mayor of Paris, was executed during the French Revolution and for what reason?
6. He was an editor of Astronomical Journal and first estimated the mass of Titan. Who was it?
7. Who introduced the Julian Day Calendar and for what purpose?
8. The amateur astronomer who identified the Crab Nebulae and also observed a rare occultation of a planet by another planet. Who was this and when did he observe the phenomenon?

9. Name the brewer who set up a telescope in Malta and discovered many satellites of the giant planets, among other things.
10. Where would you find Arago's ring and objects Larissa, Prometheus and Bianca?
11. Name two satellites of Uranus which are not named after characters in Shakespeare's plays.
12. Stony-iron meteorites named after a town in Pakistan?
13. Jupiter XIX is also known as?
14. Neumann bands are found emanating from which celestial objects?
15. The only celestial object to be named after a part of the human body. How did it come about?
16. In which story does the planet 'Lukash' occur and what is special about this object?
17. The object Joo2E3 was later identified as?
18. 'A neutron star found at F sharp'. What does this mean?
19. Who discovered the first of a class of objects called Damocloids and what are they?
20. Where would you find the following clusters: IRS 13 and Datura
21. 'Torcularis Septentrionalis' is better known by what name? What kind of object is it?
22. Where would you find the Eise Eisuiga Planetarium? When was it built and by whom?
23. What celestial event occurred on 8 May 1774 and what did Dutch astronomer Elco Alta predict to happen on that day?

24. A satellite (moon) of a planet was recently suspected to have rings orbiting it. Which one was it?
25. By what factor has the number of known asteroids increased since 2000?
26. On 22 March 1989, which asteroid passes earth at about twice the distance to the moon?
27. When is Apophis expected to encounter earth and would approach to what distance?
28. Where is XuYi observatory located? Its new one metre telescope has discovered how many asteroids in the past year?
29. LSST camera is expected to have how many pixels?
30. A 'naked-eye' gamma ray burst was recently detected by what telescope? When was this found? What was the estimated absolute visual magnitude? Day of the discovery coincides with what notable event?
31. What kind of astronomical event was "Vae Cas 2006"? When did it occur? What is it now attributed to?
32. A fifth planet was recently discovered orbiting which star?
33. What is GMT and how big would be its primary mirror?
34. Which spacecraft was mistaken for an asteroid in November 2007?
35. What are SDO and Ibex?
36. Which supernova remnant is located near the star lambda Centaurus and who discovered it?

37. Which supernova remnant is dubbed a 'textbook example' and why?
38. Why was the pulsar J1903+0327 in the news recently? What is unique about it?
39. The PICARD micro-satellite is scheduled to be launched for what purpose and when?
40. What is the SODISM telescope?
41. At what total solar eclipse (when and where) did Sir George Airy, among others, make first observance of 'pink protuberances' surrounding the sun?
42. Which astronomer, from his well equipped private observatory at Redhill, Surry, detected an intensely luminous filament, 60,000 km long extending over a sunspot? When was this?
43. John Flamsteed is supposed to have spotted a supernova in our galaxy. Which one was this and on what date did he observe it?
44. What are SOVAP and PREMES?
45. Which famous astronomers announced their discoveries in the following statements?
46. a. Cynthiae figures aemulator Mater Amorum
47. b. Annulo Cingitur, tenui, plano, nusquam coherente, ad ellipticam inclinato
48. And what are these discoveries?
49. What was unusual about supernova 2006 gy?
50. When was the Hubble spacecraft launched?
51. What is Dome C and where is it located?
52. What is the Kagulga spacecraft and when was it launched?

53. What are ANTARES, AGASA, HiRES and NESTOR?
54. Which is the brightest star in Cygnus?
55. In the southern hemisphere, which object is considered the closest equivalent to the pole star?
56. Which stars form the asterism of the 'summer triangle'?
57. Who discovered the Veil Nebula and when?
58. Halley, as is well known, predicted that the comet (now named after him) would return around 1758. However he died 16 years earlier. Who was the first astronomer to actually observe the predicted return of Haley's comet and on which day (in 1758) did he observe it?
59. What is SKA? When and where is it expected to come up?
60. What sort of an object is G29-38 and what did the Spitzer space telescope detect around it?
61. The summer solstice occurred on June 20th (on earth) in 2008. When did the Martian solstice occur?
62. How long would a summer season last near the Polar Regions on the planet Uranus and why?
63. The so called Tropic of Cancer is now actually misnamed. Why? What should it now be called?
64. What is the planned ATLAS telescope? Where is it proposed to be located and when?
65. What are JWST and TMT?
66. Who was denied a patent for inventing the telescope in 1608 and in which country?

67. What is an ED refractor and what are its advantages?
68. What kind of astronomical object is Parthenope? Who discovered it and why was it thus names?
69. Why are Barnard's star and Barnard's galaxy, notable objects? For what reason?
70. What are Subaru, SALT, HET and LBT and where are they located?
71. Stardust returned samples from which comet and when?
72. The Rosetta probe is heading towards which comet?
73. When was the Spitzer space telescope launched?
74. What is COROT?
75. The New Horizon Spacecraft was launched towards which planet and when? When would it reach its destination?
76. Which comet was briefly visible as a bright object in January 2007?
77. What is the WLM galaxy and who where its discoverers?
78. What is SagDEG and how is it connected with M54?
79. July 2nd, 1967 is a landmark for which branch of astronomy and why?
80. The first extrasolar planet was discovered around which object, by whom and when?
81. Where and when would you find Milkomeda?
82. Which object is known as the Sea Goat?
83. What is a UCDG and give some examples?

84. What is BeppoSAX? Why was it important and when?

85. Who discovered three supe-rnovae in one night for the fourth time? Which telescope was used and from which observatory?

86. Which is the nearest example of a starburst galaxy and how far away is it?

87. What was CGRO? When was it launched and how did it end?

88. What is Gyulbudaghian's nebula? With which star is it associated? Why was it in the news recently?

89. Who has more than hundred supernovae discoveries to his credit?

90. What is the maximum number of lunar eclipses in a year?

91. In which year in with next century would this maximum number be seen?

92. Who discovered that the second component of Mizar is a spectroscopic binary?

90. Which solar astronomer pioneered investigation of the infrared solar spectrum? He was the director of which observatory?

91. Who co-discovered with W Hiltner, the polarisation of star light and of which observatory was he the director?

92. Jupiter VIII was discovered by whom on January 27th 1908?

93. Taurus Poniatovii, was coined in honour of which king?

94. The well known rhyme 'Twinkle Twinkle little star' was composed by whom and when? Where did it appear?
95. What is COAA?
96. MACE 2006 stands for what?
97. What is the Merlin Medal? Who received it in 2007 and for what?
98. The phenomenon of 'Ashen Light' is associated with which celestial object?
99. What type of object is V1316 Cygni and why was it in the news recently?
100. Which is the most massive globular cluster in our galaxy?
101. Asteroid 216 is better known by what name?
102. Which amateur astronomer has discovered sixty five asteroids since 2002?
103. The Hamburg galaxy is also known as?
104. NGC 104 is better known as?
105. TheShapley-Sawyer classification is used for which class of astronomical objects?
106. Who received the Steavenson award for year 2007?
107. The planetary nebula GJJC1 is buried close to the centre of which globular cluster? How was it discovered and when?
108. How many planetary nebulae are known to exist in globular clusters?
109. What is the WASP project?
110. What is GRS in planetary astronomy?
111. What is the Good Lighting Award and who received it in November 2007?

112. What type of object is SS LMi? What is LMi? Why was it in news recently?
113. Who discovered the object in the question above and when?
114. The Crescent Nebula is associated with the ejecta of what type of star?
115. What type of object is the 'Blue Snowball' and it is found in which galaxy?
116. What is the Blinking Eye Nebula and where is it present?
117. NGC 7026 is better known as?
118. The total solar eclipse of August 1 2008, will have is maximum duration of 2 min 27 sec, at what location?
119. What is the lunar crater named after a 19th century botanist and selenographer?
120. Who was the sixth Astronomer Royal?
121. Who is the present Astronomer Royal for Scotland?
122. Name the present Astronomer Royal and who preceded him?
123. The feature Larrieu's Dam, occurs on which object? Who discovered it and when?
124. The Horace Dall Medal is given for what achievement and who awards it?
125. Who received the above said medal for 2007 and for what achievement?
126. What are JaFu1 and JaFu2 and where would you find them?
127. Pease 1 is what kind of object?
128. Give the full name of the 7th Astronomer Royal.

129. 'The polybius 'k' feature in foothills of Rupes Altai'- what are we talking about?
130. Iota Boo is what kind of object?
131. What is Geminga and why is it called so?
132. Whom did Newton succeed as Lucasian professor? Who is the present Lucasian professor?
133. The name 'Nemesis' was proposed for what kind of objects and why?
134. What is HATNET, and what is HAP-P-76?
135. What type of object is HD 189733b? Why was it in the news recently?
136. What is PG 1159? What is PG?
137. What is the NICMOS camera and where would you find it?
138. What are NEO's?
139. What was Lake Cheko recently associated with?
140. Sikhote-Alin is what kind of phenomenon and when did it occur?
141. The Arecibo telescope sent a coded message towards which object and when?
142. Dactyl is the satellite of which object?
143. What is the Lockman Hole?
144. What kind of an object is Dhofar 961?
145. Baade's window refers to what kind of phenomenon?
146. In which of Gulliver's travels did Jonathan Swift make a remarkable prediction (in the context of astronomy)? What was the prediction and what was significant about it?
147. Where would the 'Shorty' crater be found?

148. What is the inclination of the moon's rotational axis and what is significance does it have for the lunar poles?
149. What are TNO's and KBO's?
150. What are Centaurs? Give examples.
151. Where are the Kirkwood gaps to be found?
152. The Baldwin effect crops up in what class of objects and what is the effect?
153. 'Hayabusa to soon encounter Itokowa'- what is being talked about?
154. Which is the largest moon in the solar system to orbit its primary in a retrograde direction?
155. Which object was detected on New Years Day 1801 and by whom?
156. Which elements of the periodic table are named after the moon and the earth?
157. Two transuranic elements are named after which planets?
158. Where do we find the Lakshmi Plateau and the Maxwell mountains?
159. What objects are Cubewanos?
160. Thalassia orbits which objects?
161. The moon is slowly receding from the earth. Is it true? If so by how much and why?
162. When was Pluto discovered and by whom?
163. The planet Vulcan was proposed and never discovered. Explain.
164. The distance to which star was first estimated by the parallax method?
165. Who observed the first white dwarf and when?

166. Neutron stars could be remnants of supernova explosion. This was first suggested by whom and when?

167. Which astronomer first drew attention to the certain presence of vast amounts of dark matter? In which objects did he infer their presence?

168. Pulsars were first identified by whom? And in which year?

169. Who formally introduced the term blackhole in astronomical literature?

170. In what context did the star dubbed S2 become important?

171. Proxima Centauri, the nearest star from earth (apart from the sun!) has what luminosity and mass?

172. Which is the only asteroid that can be seen by the naked eye?

173. Who discovered the asteroid mentioned in the previous question?

174. Name a famous short story by Asimov which has the above asteroid in its title.

175. Which were the first four asteroids to be discovered?

176. Where would you find Archer's cluster?

177. What type of object is Hodge 301 and where is it located?

178. How many objects were listed in the Messier catalogue?

179. What is special about S Doradus?

180. The Humonculus nebula surrounds which star?

181. What are roAp stars?
182. What unit is the jansky and after whom is it named?
183. What are CME and TRACE? With which object are they associated?
184. Name five satellites of Uranus which are all named after characters in Shakespeare's play 'The Tempest'.
185. Which planet has its rotational period longer than its orbital period around the sun?
186. The Messenger spacecraft is studying which planet?
187. The gravitational tidal force between two objects is inversely proportional to what power of the distance between them?
188. The earth's rotation is slowing down by how many seconds in a million years owing to the moon's tidal drag?
189. How much tidal power is released in this slow down of the earth's rotation?
190. Which is the largest of the four Galilean moons of Jupiter?
191. Iapaetus is the satellite of which planet? What is strange about it?
192. Why was the satellite Enceladus in the news recently?
193. Who proposed the nebular hypothesis for the origin of the solar system?
194. What is the Roche limit?
195. Which European artist made a painting of Halley's Comet during its passage in 1301?

196. Name the spacecrafts which had an encounter with Halley's Comet in 1986?

197. Which spacecraft first photographed the unseen side of the moon and when?

198. Which astronomer expected the moon to be covered with a thick layer of moondust into which astronauts could sink?

199. A C Clark wrote a story using the above theory. Name the book.

200. Which spacecraft after softlanding on the moon refuted the above hypothesis? When was this?

201. Name at least two other influential theories for which the astronomer in Q.198 is known for.

202. Pluto is now known to have three satellites. Name them.

203. What is Gould's belt?

204. Name the last two astronauts to walk on the moon's surface.

205. When was this and what was the spacecraft?

206. For a given mass, what is the ratio of the energies required for escape from the gravity of the earth and the moon?

207. At the distance of Neptune, thirty times farther away from the sun than the earth, what would be the solar power on an area of one square metre?

208. Plaskett's star is a binary with a period of fourteen years? How massive are these stars compared to the sun?

209. Sirius has a mass 2.3 times solar mass and a radius 1.6 times that of the sun. Is the average density of Sirius smaller or greater than water?

210. What type of object gives rise to a Type Ia supernova?
211. What is the MEarth project?
212. What is the object Tres-4, recently discovered and what is strange about it?
213. Which astronomer has named his daughters Stella and Aurora?
214. Why is HD 209458b unusual?
215. What is the CFBD survey?
216. The L and T spectral class refers to what object?
217. What type of star is Polaris?
218. What do you understand by the so called 'Pioneer Anomaly'?
219. What is the VISTA telescope?
220. What is NOAO?
221. Which is supposedly the smallest astronomical satellite launched? What was the purpose?
222. Where is the Schickard Crater and what is odd about it?
223. On July 30, 2007, what notable phenomenon occurred in the Uranian system (that is the system of the planet Uranus and its satellites)?
224. What is HD which often occurs with a number (specifying an object) such as HD189733, etc?
225. In the previous question, when was the work done and who funded it?
226. What are MK luminosity classes? What does MK stand for?
227. MKK atlas is named after whom?

228. Who was Giambattista Riccioli and what was his contribution to lunar astronomy? When did he do his work?

229. On which object would you find the Wargentin crater and what is odd about it?

230. What is the longest duration of a total solar eclipse?

231. Apollo eleven landed on which region of the moon?

232. What is M87 and what is the estimated mass of its central black hole?

233. In which cluster is the object in the previous question located?

234. What is IMB? It detected what type of radiation and from which object in 1987?

235. What is a core collapse supernova? What type of supernova are they known as?

236. The radioactive decay of which elements is supposed to power the light curve of a Type Ia supernova? What mass of these elements is produced?

237. What type of object is expected to produce Type Ic supernova?

238. The supernova seen in our galaxy in the year 1604 is associated with which astronomer?

239. What type of supernova was the one in the previous question?

240. The brightest naked eye supernova was seen in...

241. What is a 'Moreton Wave'?

242. Who discovered it (Q.241) and when?

243. Which astronomer, on April 7, 2008, discovered his hundredth asteroid? What is the name of the object?

244. The astronomer in Q.243, had discovered how many asteroids two days earlier? From which observatory was the discovery made?

245. Which space mission was launched to observe the solar north and south poles?

246. What is the AMIGA project?

247. What is the SBF technique for distance measurements and for what object is it used?

248. What was the unusual discovery made recently using the cluster Cl J0152.7-1357?

249. What are CG's?

250. Which is the largest object known to have passed within two lunar-earth distance, from the earth?

251. Who first observed the object in question 250?

252. What happened to this object subsequently?

253. What is a Plutino?

254. On which object would you find 'Moustaches' and what are they?

255. How many binary asteroids are now known?

256. Who made the first photographic discovery of an asteroid and when?

257. What was the asteroid in Q.256?

258. Which was the first inner Oort cloud object to be discovered and when?

259. The hundredth asteroid was given what name and when was it discovered?

260. The hundred thousandth asteroid was called what name and when was it discovered?

261. What type of object is 'Hanny's Voorwerp'? Why was it so named?
262. Which galaxy is associated with the object in Q.261?
263. What toxic chemical did Phoenix probe find on Mars recently?
264. What is the LONEOS project?
265. What is the galaxy zoo project?
266. What object was named Britastra, by whom and to commemorate what?
267. The occultation of which star by Titan was widely observed?
268. What is 'gibbous moon'?
269. What type of object is Orcus?
270. What object is Myriostos?
271. Why was Hidalgo a significant object?
272. At what angle is the sun's rotational axis inclined?
273. Which object has been assigned the number 134340?
274. Which TNO is highly elongated and has at least two moons?
275. How many objects are known to have (approached earth) passed within the moon's orbit?
276. What is NEOCP?
277. What is the OGLE project?
278. M51 is popularly known as?
279. Swan bands are seen in what class of objects?
280. What was the temperature hundred seconds after the universe started expanding?
281. Who wrote the book 'The First Three Minutes'?

282. What is the significance of the first three minutes in the context of cosmology?

283. A volume of one cubic metre in intergalactic space has how many cosmic microwave background photons? How many protons are there?

284. The total solar eclipse of 16th April 1178 B.C is supposed to have been witnessed by whom in whose epic poem?

285. Which spacecraft first contacted water ice on Mars?

286. Which well known space mission finally ran out on power in July 2008? How many years did it spend in orbit, studying what object?

287. What was unique about the mission in Q.296?

288. The Ivuna meteorite fell in which country and in which year?

289. Mizar and Alcor belong to which group of stars?

290. The Tandem mission is being planned by which agency and for what purpose?

291. What is IUE? When was it launched and how many years did it operate?

292. Fe XXVI is similar to which atom?

293. What is the highest H-atom n level one would expect to find on the sun?

294. The Wilson-Bappu effect enables what quantity to be estimated and for what astronomical objects?

295. What transitions are called 'Hnb'?

296. What unusual phenomenon has been observed from MCW 349?

297. The Messenger mission is an acronym for what?

298. What are the next flyby's of Messenger past Mercury?
299. E7 and Sd are what type of galaxies?
300. Give an example of each type in Q.299.
301. The Nozomi spacecraft was scheduled to reach what planet and when? What happened to it?
302. Celestis Inc. of Houston, Texas funded which project in 1999?
303. 'Superluminal motion' was first observed in visible light from which object?
304. How far away is the Sculptor cluster of galaxies?
305. NGC 1143 is what kind of object?
306. What are LSB galaxies?
307. Give an example of the object in Q.306.
308. On which day of the year is the earth usually at perihelion?
309. What effect is used to invoke that IGM was reionized at a redshift of six?
310. What is the MARS instrument and on which telescope is it used?
311. Give an example of a quasar observed with MARS.
312. What type of object is epsilon Indi B? When was it discovered?
313. What is the mass estimated for epsilon Indi B?
314. Which telescope was used in getting IR image?
315. The Pleiades is also known as?
316. The Cassini division separates what objects?
317. What are BAT 99-2 and BAT 99-49?
318. Malcolm Coe, et al, using RXTE found 36 binary systems in 3 years. Where did they find these?

319. What is RXTE?
320. Which is the geologically most active object on the solar system?
321. The Janus hot spots occur in which object?
322. Which elliptical galaxy was (until recently) thought as lacking a dark matter halo?
323. What are solar tadpoles?
324. What is the 2MASS?
325. Which telescopes where used in Q.334?
326. When did the above survey begin and how many images were mapped?
327. Where would you find the Arabia Moisture region?
328. X ray 'Wakes' of high speed galaxies were observed from Chandra in which object?
329. What is TDRS-1?
330. Triple star system Omicron 2 Eridani is composed of what object?
331. In Q.330, what is the period of the white dwarf around the red dwarf binary?
332. Constellation Lepus and Pavonnis refer to which animal?
333. How long does a typical O type star last?
334. What is special about Iron-60?
335. Where are the galaxy's first stars thought to have formed?
336. Cooling models applied to white dwarfs in M4 give what age for the Halo?
337. As far as our galaxy is concerned, which is older: the thick disc or the thin disc?
338. What is CMD?

339. What is JAXA?
340. What are 'Spirit' and 'opportunity'?
341. What is the orbital period of the Sirius binary?
342. The Plaskett star binary has what orbital period?
343. What are CaIs?
344. Where are the above found?
345. Efremovka is what type of object?
346. Evidence for what trans-uranium isotope has been found in meteorites?
347. A red shift of z =1 correspond to what period back in time?
348. NGC 6240 is special. Why?
349. An ultraluminous x-ray source in the galaxy M82 is considered as evidence for what type of object?
350. What was the home planet of 'superman'?
351. What was supposed to be the surface gravity of planet in Q.360?
352. What is CELT?
353. What is GSMT?
354. The adaptive optics for CELT would use how many lasers?
355. What is IC 4997 and why is it interesting?
356. What is the origin of the word 'cubewano' for a class of KBO's?
357. Where is ALMA located and what does it stand for?
358. The first stars are believed to have formed when the universe was how old?
359. How many young stars have formed in the Orion nebula?

360. Till 2003, NASA operated a liquid mirror telescope (LMT) (3 metres). Where was it located?

361. What liquid is used in LMT?

362. Why is the Pilot star so called?

363. How luminous is the object in Q.362?

364. What is LBV 1806-20? What is notable about it?

365. Who claimed that the object in Q.364 is the most luminous star?

366. What does AAS stand for?

367. What force is expected to dominate the above LBV?

368. Why was the above object not discovered earlier?

369. Name a more familiar star of comparable luminosity in the Milky Way.

370. What is C153 and how is it connected to Abel 2125?

371. From Q.370, what observation is possibly explained?

372. What is R136a?

373. What is INTEGRAL?

374. Who launched INTEGRAL and when?

375. What does it consist of?

376. What energies does it probe?

377. What is HEAO-3?

378. First image of the Gamma-ray sky was obtained by?

379. What is SIGMA?

380. What is the 1.8 MeV line?

381. What gamma ray lines were detected from SN 1987A?

382. What is the COS-B satellite?

383. Which radioactive isotope is expected to still power SN 1987A?

384. Which asteroid has the smallest, fastest orbit?

385. The distance of which star was first estimated? When and by whom?

386. When did Doppler discover the effect that goes by his name?

387. Who proposed the 'nebular hypothesis' for the origin of the solar system?

388. Who first proposed a dense hot origin for the universe?

389. Who mathematically proposed the idea of a homogenous and isotropic universe?

390. Who got the Nobel Prize in 2006 for their work in cosmology?

391. Who got the Nobel Prize for the discovery of the first pulsar?

392. For what discovery in astrophysics did Hans Bethe get the Nobel Prize and when?

393. Who first reported the discovery of redshift in the spectra of galaxy?

394. Who introduced spectroscopy to chemistry, followed by identification of elements on the sun?

395. What is the proportion of heavy hydrogen (deuterium) to hydrogen in the universe?

396. Which is the heaviest element expected to be produced in the early stage of the hot dense universe?

397. In Q.396, what is the fraction of Lithium as compared to hydrogen?

398. About what fraction of the hydrogen is expected to be converted into helium in the early phase of the 'big bang'?

399. The star Zubeneschamali is also called as?

400. What is odd about the star in Q.399?

401. What is project Phoenix?

402. From what distance can the equipment in Q.401 detect signals?

403. What is S16?

404. When they are at minimum brightness, what substances are supposed to cause Mira variable to decrease to 0.1 percent in brightness?

405. Which white dwarf is expected to explode, sometime in the next half a million years?

406. What is Chi Cygni?

407. What is Sakurai's object?

408. In its full twelve days of operation how many gamma ray bursts did the Swift satellite discover?

409. What is Konus-Wind?

410. What was detected reflected off the moon by Konus-Wind?

411. When will Sirius B be furthest from Sirius A?

412. The Spitzer Space Telescope has revealed what object orbiting Vega and at what distance?

413. A 'haul' of twenty one millisecond pulsars was found in which source and using which telescope?
414. What are UHECR's?
415. KBO, 2002 LM60 is also called what?
416. Who discovered the x-ray background with a pioneering rocket experiment and when?
417. Who got a share of the 2002 Nobel Prize in physics for work on solar neutrinos?
418. Who were the other people who shared the prize in 2002?
419. On which object would you find the feature Tvashtar Catena?
420. Which spacecraft passed just 300 km above the feature in Q.419 and when?
421. When was the MIR space station launched? How and when did it stop functioning?
422. When was the Mars Odyssey spacecraft launched?
423. What nucleosynthetic process are merging neutron stars expected to enhance?
424. Where is the WIYN observatory located?
425. Where is the Pele Volcane?
426. M104 is better known as what?
427. What is M44 is called as?
428. What is Albireo?
429. An example of a galaxy not known to have a black hole at its centre?
430. What was active region 9393?
431. What is the highest designation solar flare?
432. What is the MOON project?
433. What are called HADS?

434. What are SX Phoenicis stars?

435. What type of star is YZ Boo?

436. In Q.435, who made first photographic observations of the star and when?

437. What is CGRT? Where is it located?

438. What is the Castor project?

439. How many objects did Castor track and over what period?

440. What is Sogulu? Which object was supposed by whom to be made of this material?

441. The Rockefeller reflector is located in which observatory?

442. When was the Boyden Observatory established and where?

443. Who funded the Boyden Observatory?

444. Who decided on S America for a suitable site?

445. Why was the Baker-Schmid telescope called the ADH?

446. What is 'glory effect'?

447. What is the Chant Medal?

448. To whom is the medal (in Q.447) given and who was the recent recipient?

449. Who was the first recipient?

450. What is WMAP?

451. The Planck satellite is to be launched when and for what purpose?

452. Radio-astronomer Martine Ryle got the Nobel Prize when and for what?

453. What was the BOOMERANG experiment?

454. What is a hypernova?

455. What object is responsible for a nova outburst?

456. 'Vulpecula' is associated with which animal?
457. The nearest star to the sun, Proxima Centauri is what type of star?
458. Procyon is what type of star?
459. The companion of Procyon is what type of star?
460. What was the type of star which exploded as SN1987a?
461. How far away did the event (in Q.460) occur?
462. What does SS stand for, for example in SS-433?
463. 'Cytherium' refers to which planet?
464. When did comet Shoemaker-Levy collide with Jupiter?
465. What was the size of Fragment G, in Q.475?
466. How much energy was released in the collision of Fragment G with Jupiter?
467. In terms of the Hiroshima bomb, how much does this correspond to?
468. What is the velocity with which an object like fragment G would hit Jupiter?
469. What are Bok globules?
470. How old is the star HE 1523-0901?
471. How do we know it is that old?
472. Which was the first planetary nebula to be discovered?
473. Who discovered it and when?
474. In which constellation is it located?
475. M11 is commonly known as what?
476. Who gave the name to M11 and when?
477. What is the MAMBO-2 instrument?
478. What is LABOCA device?

479. What is the APEX collaboration?
480. Where would you find the Mohorovic discontinuity?
481. Where is the Gutenberg discontinuity?
482. What do the terms FWHM and PSF stand for in astronomical data?
483. Name the spiral arms of the Milky Way galaxy.
484. Does the earth-moon distance remain same?
485. In Q.484, how do we know this?
486. What is the reason for the increase in earth-moon distance?
487. Will the moon's distance from earth continue to increase indefinitely?
488. Which Dark Nebula is known as the 'Inkspot'?
489. Which eclipsing binary star has the longest known period?
490. What new name has the IAU decided for Trans-Neptunian dwarf planets?
491. What is the fastest spinning natural body known now in the solar system?
492. Who discovered the object in Q.491?
493. Which is the most recent Milky Way supernova?
494. On which red dwarf star a great big flare, outshining the star several times, erupted on April 25, 2008?
495. In Q.494, what is the star's rotation period and magnetic field?
496. Which neutron star emits almost all its energy as gamma rays?
497. On which latitude on Mars did Phoenix spacecraft land?

498. In Q.507, the first excavations of Martian soil were on a trench called?
499. The lightest exoplanet found has what mass?
500. Where would you find the Victoria Crater?
501. Why was the crater (in Q.510) in the news recently?
502. Where is the Orwell Park Observatory?
503. What is AKR?
504. What does 'GOALS' stand for?
505. What did the Spitzer space telescope discover about Omega Centauri?
506. What does Capriconus mean in Latin?
507. Why is Capriconus also called 'sea goat'?
508. Where did the Iuana meteorite fall?
509. Where would you find the Hadley Rille?
510. Which astronomer discovered that the star Algol varied in brightness?
511. What is ASCOM?
512. Where would you find Borealis Basin?
513. What are Iridium flares?
514. What is 'Brownleeite'?
515. Who is Brownleeite named after?
516. Where did the dust particle in Brownleeite supposedly originate?
517. When was the comet Brownleeite discovered?
518. What are sundogs called in German?
519. Which German poet wrote "Drei Sonnen sah ich Himmel Stelin", and what does it mean?
520. What is GOODS?
521. Where is the Adviar impact crater?
522. Cerberus Fossae refers to what?

523. What does Fossae mean?
524. Where would you find the snow white trench?
525. Who was Ferdinand Verbiost?
526. What was Verbiost noted for?
527. Who preceded Verbiost at the Emperor's court?
528. What else did he install in the Beijing Observatory?
529. Which Japanese artist pictured Verbiost in a woodcut?
530. Verbiost also doubled as what?
531. Which astronomer had an ensemble of massive astronomy instrument on the island of Hven?
532. Where is the Louth crater?
533. On which object would you have the Stickney crater?
534. What is MOLA?
535. Which satellite identified Gemina as a spinning neutron star and how?
536. What type of star is Epsilon Aurigae?
537. Which star was recently found to have a "hot Neptune" and a 'super-earth' orbiting it?
538. Multiple planets have been identified around how many stars till date?
539. Why is Alpha Centauri- B considered suitable for using the radial velocity method to identify planets around it?
540. What fraction of sun-like stars are currently known to have giant planets?
541. What is Ara 1520 and who discovered it?
542. Who is credited with the discovery of cosmic rays?

543. Which is the largest satellite in the solar system?
544. How much larger in diameter is it than Titan, the second largest satellite?
545. Jupiter XVIII is also known as what?
546. Where would you find the "Galileo Regio"?
547. The diffuse innermost ring of Neptune is called what?
548. GALEX stands for what and what does it deal with?
549. What is the SAGE project?
550. Neptune VI is also known as what?
551. What is the average strength of the galactic magnetic field?
552. The OH maser emission in astrophysical source is prominent as what frequency?
553. Name other types of masers in astronomical sources.
554. What comet known earlier as 1983 VII came within 4 million kilometres of earth? Which satellite discovered it?
555. IRAS stands for what?
556. Which asteroid did the English astronomer John Hind find in 1847?
557. Jupiter XXIV discovered in 200 is also known as what?
558. The 4.2 m William Herschel Telescope (WHT) is located where?
559. To which telescope group does WHT belong? Who owns it?
560. What is JKT?

561. Which satellite is dubbed ISEE?
562. What is IGM?
563. Which double star system is dubbed "Pulcherrina" and why?
564. What is NTT?
565. Where is La Silla located?
566. Which constellation represents a hare?
567. Who gave the satellite Titan its name?
568. Which astronomer deduced that Titan has an atmosphere and when?
569. Who first detected the presence of methane on Titan?
570. The surface atmospheric density on Titan is how many times that of sea-level air on earth?
571. How long is a day on Titan?
572. What is unique about the object 6R 10D B9?
573. How many orbits did the object in Q.572 make around earth?
574. What does 'Wolf number' refer to?
575. RGO stand for what?
576. What is the 'harvest moon'?
577. What does REO stand for?
578. Which astronomer first measured the speed of light?
579. The Roud-Sastry classification scheme applies to what class of objects?
580. Which astronomical organisation was formed in 1831?
581. What effect deals with an apparent anisotropy in the expansion of the universe over scales of about 50Mpc?

582. The five kilometre radio telescope was renamed as?

583. 'Regmaglypts' are found on what kind of objects?

584. Alpha Leonis is better known as what?

585. Who was the first person to build a reflector telescope and when?

586. What are RRAB and RRC stars?

587. The elements such as gold or uranium are expected to be produced in what nuclear process and where?

588. The 'Victoria Rupes' is a feature on which planet?

589. SAMPEX stands for what?

590. The triple alpha process, produces what element? It also called what?

591. Name the Japanese space probe (to Halley's Comet) whose name mans 'pioneer' in Japanese.

592. Who showed in 1866 that the Perseid meteor shower was associated with comet Swift-Tuttle?

593. Who first obtained UV photographs of a solar eclipse?

594. What are mascons?

595. What are the consequences of mascons for lunar flights?

596. What is selenodesy?

597. What is GRACE spacecraft and what was the purpose?

598. When was the mission in Q.597 launched?

599. Mneme is what object?

600. Which star has the name that means girdle in Arabic?

601. What is the proposed GRAIL mission?
602. Which star has the name that means belt in Arabic?
603. What are Kaguya, Ouna and Okina?
604. For the moon, what is the corresponding term for geoid?
605. What is MCAO?
606. How does the system in Q.606 work?
607. What is E-ELT?
608. What is the proposed aperture of E-ELT?
609. Which is the 'driest' impact basin in the moon?
610. What is meant by 'driest' in Q.620?
611. Where would you find the crater Bullialdus?
612. The Archimedes crater on the moon is found within which basin?
613. How did the distant minor planet Sedna get its name?
614. Which satellite's data indicated accelerated melting of Greenland's ice sheets?
615. Which was the first object in the Messier catalogue?
616. Who appended the name to the above object?
617. What is BATSE?
618. Where is the Tidbinbila radio telescope?
619. Which 13th century Persian astronomical writer wrote that people tested their eyesight by seeing which star?
620. Which is brighter, Mizar or Alcor?
621. A redshift of two corresponds to what distance from our galaxy?

622. In an accelerating universe, dominated by dark energy, which may be the only galaxy visible to astronomers, say in twenty billion years?

623. Which meteorite fell on October 1992, damaging a parked car?

624. The star Alpheratz is also known as?

625. Sylvia and Hilda belong to which class of asteroids?

626. Pedestal craters are chiefly found on?

627. Features prominent in the spectrum of carbon star and lying at 1207 nanometre are called what?

628. Seventh closest satellite of Uranus, named after a character in 'Merchant of Venice' is?

629. The binary star Gamma Virginis is also called?

630. A sunspot without a penumbra is called?

631. The Poynting-Robertson effect is due to?

632. Which meteorite fall was first photographed by a camera network?

633. Which journal was founded in 1889 and now published monthly?

634. (1862) Apollo is perhaps the only one in this category. Which one?

635. QSS stands for what?

636. Where would you find the 18 km crater Yuty?

637. What is RATAN-600?

638. Name the only one member of the R-class asteroid?

639. NGC 7009 is better known as?

640. What is a trischiefspiegler?

641. Eunomia and Amphitrite belong to what class of objects?
642. What is SCT?
643. What is STScI?
644. Name the space probe whose name means 'comet' in Japanese.
645. Solar activity in which material ascends then descents vertically?
646. Name the outer moon of Uranus named after a witch in 'The Tempest'.
647. Curve connecting regions in comet dust tail containing identical size particles is called what?
648. What is the feature on Mars, whose name means 'the great sand-bank' in Greek?
649. Name the element number 43, not occurring in nature, but found in some carbon stars.
650. Yet another unstable element, which is a lanthanide.
651. Elements such as boron or beryllium are produced by what type of nuclear process?
652. Name the alloy of tin and copper used earlier for telescope mirrors.
653. Which is the hypothetical rapidly spinning supermassive star envisaged to be a scaled up version of a pulsar?
654. What is the reflection nebula IC 2220 (shaped like a tankard) also called as?
655. Which asteroid passed within twenty million km of earth in 1972?
656. What is galaxy M33 also called as?

657. The Trumpler classification applies to what objects?
658. The observed correlation between the width of the 21 cm hydrogen line from spiral galaxies and their absolute magnitude is called what?
659. A relation similar to that in Q.669, bit applicable to elliptical galaxies is called what?
660. The tuning fork diagram depicts what?
661. Which comet first seen in 1858, was rediscovered in 1907 and again in 1951?
662. Which catalogue was published in 1997 from observations compiled from Hipparcos satellite?
663. The star Gamma, Zeta, Eta and Pi Aquarii forms what?
664. When and where was the WFPC 2 installed?
665. The Servicing Mission 3B was when and what was installed?
666. What is NICMOS?
667. When was NICMOS installed?
668. What is HST?
669. HST directly imaged which star's disc in 1996?
670. What is CHARA?
671. Which was the first main-sequence star other than the sun to be imaged by CHARA and when?
672. How fast does Altair spin?
673. How much faster is this than the sun?
674. What is Altair's angular size as seen form earth?
675. How many supernovae did the IAU officially name in 2007?
676. If one can detect (the brightest) type Ia supernovae out to ten billion light years how many

observable supernovae would you expect per second?

677. So in Q.676 what fraction of potentially observable events were seen?

678. The July 22, 2028 total solar eclipse would be what duration and would be seen mostly from where?

679. The August 21, 2017 and April 8, 2026 total eclipses would be both visible from where?

680. Which three total solar eclipses, after 2025, in the 21st century would exceed six minute totality?

681. Who awards the Nier Prize and who was a recent recipient?

682. Which asteroid is named after the Nier Prize winner in Q.681?

683. Wadhwa is the head of which institute?

684. Asteroid 944 was discovered in 1920 by whom and what is it called?

685. What is special about the object in Q.684?

686. Ionosphere E layer is also called what?

687. Name the smaller satellite of Saturn which shares same orbit as much larger satellite Dione.

688. What effect has been used to measure the weak magnetic fields of solar prominences?

689. What are the typical fields in solar prominences?

690. Hard X-rays correspond to what energies?

691. What is HCO and when was it founded?

692. In which year is Halley's Comet expected to return?
693. Which was the first Japanese X-ray astronomy satellite and when was it launched?
694. Japanese radio astronomy satellite, also called Haruka, means what?
695. Who introduced the Gregorian calendar and when?
696. What is GBT?
697. The star beta-Centauri (Agena) is also called what?
698. Site of the Apollo 15 moon landing in 1971 is called what?
699. The great Dark Spot is found on which planet?
700. Apart from Halley, which other comet did the space probe Giotto pass by?
701. A diagram of the energy levels for a given atom or ion consulted by astronomers is called what?
702. The part of a Sundial that casts shadow is called?
703. Which is the Japanese X-ray satellite which was known as Astro-C before launch?
704. What does GMC stand for?
705. Soviet X-ray and gamma ray astronomy satellite launched in December 89
706. What is GSFC and when was it founded?
707. What is GTC and where is it located?
708. Who in 1668 used the inverse square law for brightness of stars to estimate the distance to Sirius?
709. R Leporis is what kind of star and who discovered it?

710. Name the world's largest known meteorite found in 1920.
711. What obscures the star 'Eta Carina'?
712. Who in 1639 observed a transit of Venus that he had predicted from Kepler's tables?
713. 'Ceraunius Fossae' is found where and is an example of what?
714. A spacecraft trajectory from one orbit to another which involves minimal energy expenditure is called what?
715. Radius of a galaxy at which surface brightness is 26 mag/sq.arcsecond is called what?
716. The SAS-3 satellite discovered this white dwarf was a strong emitter of soft X-rays. Name the white dwarf.
717. IQSY stands for what?
718. Celestial hydroxyl and water megamasers generate what power?
719. The third 8.2 m unit telescope of VLT in Chile.
720. What does the name in Q.719 mean?
721. Which American astronomer proposed in 1954 that the surface of Venus is covered with water?
722. Without knowing the sun's mass or radius or its distance from earth, what property of the sun did Newton deduce and using what law?
723. The Herschel crater, 130 km across is found on which body?
724. Who recorded the first spectrum of Vega in 1872?
725. Which is the scale for roughly estimating darkness of total lunar eclipse?

726. Name the sixth largest main belt asteroid.
727. Which space probe will use ion propulsion to travel to which asteroid?
728. Third closest satellite of Neptune, orbiting just inside what?
729. What does the letter D signify on Morgan's galaxy classification?
730. What are CD galaxies?
731. Model of universe which are homogenous but not isotropic.
732. The end state of a closed universe (density greater than critical density) is described as what?
733. The critical density which determines whether the universe is closed or open depends on which cosmic parameter?
734. For a Hubble constant of 100km/s/Mpc what is the critical density?
735. ESA-Japanese mission to planet Mercury planned for 2010 is called?
736. Star Gamma Orionis is also called what?
737. First to notice a signal from what proved to be a pulsar in August 1967?
738. What was this first Pulsar designated as?
739. The emission nebula Sharpless Z-276 is often called what?
740. On what object would you find the Maunder's crater?
741. What is PHA?
742. How many PHA's are listed?

743. What is particular about the PHA's 2007 RR_9 and 6344P-L?

744. How close did comet Hyskutake come to earth and when?

745. Who sketched comet Halley at its 1607 visit?

746. Which king struck some bronze coins depicting a foreboding hippeus (horse) comet?

747. During which Roman emperor's reign in AD 66 (February) Halley's Comet was witnessed?

748. Which 2000 year old manuscript contains 29 drawings of comets?

749. NGC 3603 is noted for what?

750. Cauchy crater is a small crater on what?

751. M44 is also called what?

752. The sixth asteroid to be discovered was which and by whom and when?

753. The object in Q. 752 is named after whom?

754. Which is supposed to be Mars's youngest outflow channel?

755. What is Kemble's Cascade?

756. Which Greek scholar first estimated the circumference of the earth correctly?

757. Name the five elements with the largest cosmic abundance.

758. What is an Einstein cross?

759. Who was the ninth astronomer royal and in what mission did he play a key role?

760. What is a DC white dwarf?

761. Majority of white dwarfs are of type DA. What are they?

762. Where would you find the Schiaparelli Dorsum and what is it?

763. What is a dMe star?

764. Who invented the heliometer and developed the achromatic lens?

765. What model did Whipple propose for a comet nucleus?

766. What is Dicke receiver?

767. Where would you find the Diana Chasma?

768. What are DIB's?

769. A feature previously known as Agathodarmon is found on which planet?

770. Who invented the coronagraph?

771. Eskimo Nebula is also called what?

772. When was the Clementine spacecraft put into lunar orbit and which asteroid was it later supposed to rendezvous with?

773. Why was the rendezvous mission abandoned?

774. Dysnomia is the name of the moon of which object?

775. Which was the first asteroid to be discovered from America? Who discovered it and when?

776. Ophelia and Cordelia are what types of objects and what role do they play?

777. Where would you find the Encke gap?

778. What is Arthusa? What is odd about it? Who discovered it and when?

779. Canopus, next to Sirius appears brightest in the night sky. It has often be used as a guidance star for spacecraft. In Greek legend who was Canopus?

780. What are Kallichore and Kaluke?
781. What is unusual about Janus and Epimethus?
782. Where would you find the Keeler gap?
783. Which is the fourth largest asteroid? Who discovered it and when?
784. What is the Kleinmann-Low nebula?
785. Which astronomer was supposed to be the first to measure the Doppler shift and for which object and when?
786. The name of which star is Arabic for 'sheep'?
787. Where would you find the McMath-Pierce Solar Telescope?
788. Where and what is the Mayall telescope?
789. Who first discovered Helium on the sun?
790. Which star's name is Arabic for 'fish's mouth'?
791. LINEAR is what project?
792. What constellation stands for 'LIZARD'?
793. Who introduced the 'LIZARD'?
794. What is Henrietta Leavitt noted for?
795. Kappa Crucis is the brightest star of what?
796. The 'KIDS' stand for which objects?
797. Who founded the Lowell observatory and what telescope does it have?
798. KSC stand for?
799. What is Francisco?
800. What is Kiviuk and what is unusual about it?
801. What is the Hungaria group?
802. The footprint Nebula is also called?
803. Where did Lunokhold 2 land and when? What distance did it cover?
804. What is the LBT and where is it located?

805. What is KAO?
806. What is odd about 216 Kleopatra? Who discovered it?
807. Where are the Trojan asteroids located?
808. When are L4 and L5 stable?
809. Which sequence began on November 9, 1853?
810. What is the full name of the world's largest fully steerable radio telescope?
811. Which dwarf galaxy is presently being cannibalised by the Milky Way?
812. Where does the south galactic pole lie?
813. What is the orbital speed of the sun around the galactic centre?
814. What is IOTA?
815. Which English amateur astronomer's pioneering sky atlas was published posthumously?
816. Jupiter XXV is also known as?
817. The hour circle that passes through the celestial poles and the vernal and autumnal equinoxes is called what?
818. Alpha Eridani and Theta Eridani are better known as what?
819. Which constellation represents a foal?
820. The strength of a line in the spectrum is measured by what?
821. The Dill Faulkes educational trust provided what telescope?
822. What name is given to fine dark lines arranged in a near spiral pattern around sunspots?
823. Saturn's C ring is also called what?
824. What is CrAO?

825. The Balmer series H_a line corresponds to which Franhoffer line?

826. Where would you encounter the Butterfly diagram?

827. Which American astronomer compiled a five volume general catalogue of stars in 1936?

828. In which astronomical objects would you notice Bowen fluorescence?

829. Who developed the solar magnetograph?

830. Noise extending uniformly over a broad band frequency is called what?

831. What is Cor Caroli?

832. Which is the 'flattest' star?

833. Which bright star in the solar neighbourhood is suspected to be from another galaxy?

834. Where is the David Dunlap Observatory?

835. What type of telescope is there in the observatory in Q.846?

836. Which famous American astronomer was the Dean of the Medical faculty in New York University?

837. Who was the first to photograph the Orion Nebula?

838. Where is the Teide Observatory?

839. What are Thelxinoe and Thyone?

840. What is the difference between Taygeta and Taygete?

841. Which bright comet was discovered from Florence in 1858?

842. Which star's name means 'breast' in Arabic?

843. Which star's name means 'shoulder' in Arabic?

844. The WIYN telescope is located where and operated by?

845. The Wilson effect arises in?

846. What type of object is the Owl Nebula?

847. What are Paaliaq and Pallene?

848. What is Sirrah?

849. Where is the Pulkovo Observatory and when was it founded?

850. What object or feature is named after the Greek geometer Pythagoras?

851. What are Plerions?

852. Where is the Oschin Telescope?

853. What is an Orrery?

854. How did the name Orrery originate in Q.853?

855. Which was the largest asteroid (>300 km in diameter) discovered by Raymond Dugan in 1903?

856. What type of stars show ZrO, VO, etc. in their spectrum?

857. Which asteroid discovered from Vienna Observatory in 1911 was named after a benefactor?

858. What is the Geotail probe?

859. What is Stock 2 and how did the name originate?

860. Stock 5 is located close to which star?

861. Delta Cas is also known as?

862. Messier's original list stops at how many objects?

863. Who added the additional six, including M104 (Sombrero)?

864. What is a bolide?

865. What is aeronomy?

866. Which periodic comet is expected to be visible in November 2008?

867. Who discovered the comet in Q.866?
868. What is the DB gap?
869. What is JDEM and SNAP and what do they deal with?
870. What is ADEPT and when is it scheduled to be launched?
871. How many supernovae have been found in the galaxy NGC 6946 over the past ninety years?
872. Which was the latest supernova in the galaxy in Q.871?
873. What is MSL and which are its planned most tantalising destinations?
874. What is BRT?
875. What planetary event is expected on September 4, 2009?
876. Why would they not be seen from August 10 to September 4?
877. What is unusual about the sun in 2008?
878. What is the 'Novaya-Zemlya' effect?
879. How did the effect in Q.878 get its name?
880. A new Plutoid in the solar system discovered very recently has been officially named what?
881. Now how many Plutoids are there?
882. What is the origin of the name in Q.881?
883. Why was V1280 Scorpii in the news recently?
884. What is the Peony Nebula star?
885. Angrite meteorites are supposed to come from?
886. The Harran Sulci can be found on which object?
887. What does the term 'Sulci' mean in Q.886?
888. The Maxwell gap is found where?

889. What was Maxwell's contribution which motivated the naming in Q.888?

890. What does CP in front of a star's name stand for?

891. What Hirayama family of asteroids are characterised by their high orbital inclination of 15 degrees?

892. Who discovered the asteroid Maria referred to in Q.891?

893. The Malmquist bias appears in what context?

894. A star with an unusually high ratio of manganese to iron is called what?

895. In Q. 894, if the star in addition has a spectral line at 3984 Angstrom unit wavelength it is called what?

896. Where would you find the Tyre Macula and what does it refer to?

897. Lines separating regions of opposite magnetic polarity on the Sun are called what?

898. A birefringent filter is also called what?

899. Ophir Labes occurs where? What does it refer to?

900. What correction takes account of the effect of redshift on a galaxy's spectrum?

901. What eye-piece is often used in binoculars?

902. What are KREEP rocks?

903. A combined diffraction grating and prism is called what?

904. The ratio of the tidal forces exerted by the moon and the sun on the earth is in the ratio of their average densities. True or false?

905. Justify the answer in Q.904.

906. Which law did Newton use to deduce that sun's average density is just above that of water?

907. The space gamma ray telescope GLAST has recently been renamed as what?

908. What is the Wiggle-Z survey?

909. When did the survey in Q.920 begin and when is it expected to end?

910. What is SUMI?

911. What is Gleissberg Cycle?

912. What is the luminosity (in lumens/m^2) of a first magnitude star?

913. For a star just visible for the naked eye at an apparent magnitude of 6, how many photons (at 500 nm wavelength) would enter the observers' eye per second?

914. The distance between the rear lens of an eyepiece and the exit pupil is called what?

915. What are exit pupil and eye lens?

916. What is inverse Evershed flow?

917. Outward moving facular points form a so-called what?

918. What is EVN?

919. A periodic disturbance in lunar position caused by changes in its orbital eccentricity caused by sun's gravity is called what?

920. Heaviside layer is called what?

921. How many astronomical units are there in a parsec?

922. Time taken by the sun to orbit around galactic centre is called what?

923. Time taken for precession of the earth's pole (about 25,800 years) is called what?
924. What is Sarandib Planitia?
925. What is PLC?
926. Volcanoes Kilauea and Mauna Loa, Hawaii are examples of what?
927. A Newtonian reflecting telescope is an example of what type of systems?
928. What is the Bigelow Sky Survey and what is it now called?
929. Who was the fourth Astronomer Royal and in which year?
930. Who was the Astronomer Royal in the 1980's?
931. What is CARMA?
932. The Bautz-Morgan class pertains to what objects?
933. Who publishes 'Mercury' and what is it?
934. What is the Henry track?
935. Who originally employed as a janitor became a prominent astronomer?
936. Which Dutch astronomer predicted that neutral interstellar hydrogen should emit 21 cm radio waves?
937. The first meteorite fall photographed by a camera network was which?
938. Which observatory was involved in Q.937?
939. Explosive solar 'prominence' with very high velocity is called what?
940. With which star would you associate 'standstills'?
941. Which was the star atlas published in 1603 by J Bayer?

942. Which is the nearest white dwarf after the companions of Sirius and Procyon?

943. Name the M6 red dwarf, one millionth of the absolute luminosity of sun (one of the faintest), 20 light years away?

944. What is VATT?

945. The star Delta Canis Majoris an F8 supergiant is also called what?

946. What is the Yellow Void?

947. What is FIGGS?

948. If the sun were to be replaced by a black hole of nine solar masses, what would be the orbital period of the earth around the new primary?

949. If the sun were to be replaced by a neutron star of exactly same mass, what would be the earth's orbital period?

950. A number of 'orbit raising' manoeuvres are involved before Chandrayaan enters its final lunar orbit. At one stage its orbit around the earth is highly elliptical, with an apogee of 267,000 km and a perigee of 470 km. What is its orbital period around the earth for this particular orbit?

951. Chandrayaan's final circular orbit around the moon would be at an altitude of 100 km above the lunar surface. What would be its orbital period around the moon?

952. When Chandrayaan is in orbit in Q.951, a moon impact probe (MIP), to land on the lunar surface would be launched. How long would this

probe take to impact on the lunar surface from this orbit?

953. Does the answer to Q.965 incorporate any universal law relating the orbital period and the time taken for the impact?

954. Which science fiction writer invoked the above relation in celestial mechanics and in which story?

955. For a solar intensity of one kilowatt per square metre, how many photons of average wavelength 500 nm impact this area per second?

956. How far should the sun be for the intensity in Q.955 to correspond to one photon per second?

957. If one does not consider moonlight, the combined intensity of light from all stars in the night sky falling on earth is what fraction of the intensity of sunlight?

958. In Q.957, how many photons would the combined starlight intensity correspond to, per square metre per second?

959. Where would you find the instruments TEGA and MECA?

960. Expand the acronyms in Q.959.

961. Where is the Gale Crater and what is special about it?

962. The OMEGA spectrometer is placed where?

963. What are KM3NeT and LAGUNA?

964. A next generation gravitational wave antenna has been dubbed what?

965. The Miyamoto crater is located on which celestial object?

966. He 104 is what kind of object?

967. What does He stand for in Q.979?

968. How did the astronomer Henise meet his demise?

969. If our solar system were proportionally reduced so that the sun-earth distance is now one metre, how long would a year be?

970. Members of a binary star system have same mass as the sun. If their distance apart were equal to earth-sun distance, what would the binary orbital period be?

971. Due to the tidal effects of lunar gravity by how much is the 27 km circumference of Large Hadron Collider circular ring expected to stretch?

972. Which is the largest dark matter (DM) particle postulated?

973. If the axion is the DM, what is its expected mass as compared to that of the electron?

974. How is the axion expected to be detected in lab experiments?

975. Which particle is the favoured cold dark matter candidate and what is its expected mass?

976. What is the expected surface gravity (that is acceleration) on a neutron star surface?

977. If a neutron star magnetic field is dipolar and the surface field (radius of star is about 10 km) is a trillion gauss, what is the magnetic field ten thousand kilometres from the surface?

978. One cubic metre of matter in a white dwarf would have a mass of how many tons?

979. What would be the escape velocity from a white dwarf?

980. A distant object at a redshift of two would be receding away at what velocity?

981. What is Segue 1 and what is unusual about it?

982. How many short GRB's (gamma ray bursts) are listed in the BATSE catalogue in nine years of operation?

983. The GZK effect refers to what?

984. Which observatory has the 1.5m Tillinghast reflector?

985. The a and A bands of the Franhoffer spectrum are due to what?

986. MSSSO and NTT stands for what?

987. The meteorite that fell in southern France in May 1864 was found to have very substantial amount of organic compounds?

988. Which astronomer in 1845 installed a 72 inch telescope on his family grounds?

989. What do the square brackets denote in [O III] and C III]?

990. What are Hertha and Nysa families?

991. Which periodic comet with the longest period has been seen more than once and what is its period?

992. What was discovered recently around the white dwarf G2-9-28?

993. Which astronomer is supposed to have first suggested making telescope mirrors from a rotating basin of mercury?

994. Who first constructed a liquid mirror (mercury) telescope and when?

995. Who described the universe as a 'free lunch' and in what context?

996. What are V777 Her Stars?

997. Which are the hottest pulsating white dwarfs?

998. What is unusual about SN 2008 D?

999. In which galaxy did the above SN occur?

1000. "We are all in the gutter, but some of us are looking at the stars". Who said this and where?

1001. Who approved the motto, for what and when: "per ardua ad astra"? what does it mean?

□□□

Astronomy Quiz

ANSWERS

1. 2003, EL61 Asteroid (KBO) spinning with period of 3 hours
2. Star having one of the highest metal abundances, about four times that of the sun
3. It is named after Kuniko Sofue, a Japanese school girl who noticed a bright new star while looking through a window, darning socks late at night! This was the earliest detection of the bright nova CP Pup in 1942 in Japan
4. The Japanese barber Kazuaki Gomi, discovered Nova Lac (CP Lac), during a total solar eclipse on 19 June 1936
5. Jean Bailly, executed in 1793 during the French revolution
6. Dirk Broueer
7. The French scholar, Joseph Scaliger introduced the Julian Date calendar in 1582, naming it to honour his father, Julius Caesar Scaliger (1484-1558). So it has nothing to do with Julius Caesar!
8. John Bevis, observed the occultation of Mercury by Venus on 28th May 1737

9. William Lassell
10. Arago's ring around Neptune, Bianca satellite of Uranus, Larissa satellite of Neptune, Prometheus satellite of Saturn
11. Bianca, Belinda
12. Lodranite, named after the town of Lodran in Pakistan
13. Jupiter XX Taygete, Jupiter XVIII Themisto, Jupiter XXV Erinome
14. In iron meteorites
15. Coma Berenices (represents hair of the Egyptian queen Berenice)
16. In Isaac Asimov's science fiction story 'Nightfall'. It is a planet in a multistar system, its habitants witnessing unusual eclipses
17. Identified as the spent stage, SIVB, of the Apollo 12, manned lunar spaceship which on being jettisoned went into orbit around the sun
18. Fastest spinning pulsar, spinning 716 times a second, that is a frequency of 716 Hz, corresponding to F sharp in musical scale
19. Asteroid Damocles with an unusual orbit (which may cross that of earth, hence the name from 'Damocles Sword', in a few thousand years), discovered by Robert McNaught in 1991. This class of objects are called Damocloids
20. Near the galactic centre and in asteroid belt
21. Imicron Piscium, a model h-type star of magnitude +4

22. World's oldest functioning planetarium, built by an 18th century wool-comber and amateur astronomer Eise Eisinga in Netherlands in 1775
23. Rare conjunction of four planets and the moon on that day. He predicted the world would end with the celestial bodies crashing
24. Rhea
25. By a factor of nearly five
26. Ascelepius
27. April 2029, at around 30,000 km from surface of earth
28. Tieshan temple national forest, Jiangsu, China. Over 300
29. 3.3 billion
30. This was GRB 080319, which occurred on 19 March 2008. NASA's high energy SWIFT telescope alerted automated camera on ground. It reached an apparent magnitude of 5.4, so that its absolute magnitude is a fantastic -34, three million times more luminous than the brightest recent supernova (2005ap). The date coincided with the death of Arthur C Clark.
31. Spectacular microlensing magnification of an A type star on October 31st 2006
32. 55 Canceri
33. Giant Magellan Telescope; 8.2 metres
34. Rosetta
35. Solar Dynamics Observatory; Interstellar Boundary Explorer
36. G 292.0 + 1.8
37. G 292

38. It is a millisecond binary pulsar with a high orbital eccentricity; it is a rare type of system
39. PICARD microsatellite is to be launched in 2009 on sun-synchronous orbit. Named after Jean Picard, it would measure solar diameter, limb shape, oblateness to milliarcsecond accuracy and possible variation with solar activity.
40. The SODISM telescope of the PICARD mission will perform diameter measurements by imaging sun on CCD camera. Stands for solar diameter imager and surface mapper.
41. At the 1842 Turin total eclipse
42. Richard Carrington; in 1859, September 1st
43. Cas A, possibly August 16th 1680
44. It is the Solar Variability Picard Instrument and Precision Monitor Sensor
45. The discoveries were

    a. Galileo announcing that Venus has phases like the moon in 1610

    b. Christopher Huygens in 1656 announcing Saturn has a thin flat ring surrounding the planet and inclined to it
46. It is the brightest supernova ever seen, hundred times brighter than a typical supernova. Seen in September 2006
47. On April 24th 1990
48. It is a plateau in Antarctica, considered ideal for location of future telescope
49. Japan's lunar spacecraft carrying high definition video camera in orbit

50. Antarctic Muon and Neutrino Detector
51. Deneb
52. Sigma Octantis
53. Altair, Vega and Deneb
54. Sir William Herschel on 5th September 1784
55. German farmer and amateur astronomer Johann Georg Palitzsch first saw the predicted return on Christmas night in 1758.
56. World's largest radio telescope Square Kilometre Array (SKA). Either in Western Australia or South Africa. Construction expected during 2012-2020
57. It is white dwarf Spitzer space telescope detected cometary material around it.
58. June 25
59. 42 years. Its axis of rotation is nearly aligned with plane of the orbit. Orbital period of 84 years.
60. Due of shifting of earth's axis, the sun now appears in constellation Taurus.
61. It stands for Advanced Technology Large-Area Space Telescope and is a proposed 8 m telescope to be put at L2 around 2020
62. James Webb Space Telescope and Thirty Metre Telescope
63. Hans Lipperhay in Netherlands
64. Extra Low Dispersion, ED glass cuts down chromatic dispersion
65. It is an asteroid, discovered by astronomer Annibale de Gasperis in 1850. Named after the name of the Siren of the sea, who founded his

home city of Naples according to Greek mythology.

66. Barnard star (only six light years away) has the largest proper motion and Barnard's galaxy (NGC 6822) was the first object shown to lie beyond Milky Way's boundaries. Both discovered by E Barnard
67. Subaru is the Japanese 8.3 m telescope at Hawaii, SALT is South African Large Telescope (10m), HET (High Energy Telescope), LBT (Large Binocular Telescope)
68. From Wild 2 in January 2006
69. Comet Churyumov-Gerasimenko
70. August 2003
71. Convection Rotation and Planetary Transit; mission led by French Space Agency in collaboration with ESA to search for extra-solar planets and to measure astro-seismology.
72. Towards Pluto; January 2006; around 2015
73. Comet McNaught
74. Stands for Wolf-Lundmark-Melotte. Max Wolf in 1909, Knut Lundmark and Jacques Melotte independently rediscovered it in 1926
75. Sagittarius Dwarf Elliptical Galaxy, M54 resides close to its core and is the second most massive of galaxy's globular clusters
76. Gamma Ray Astronomy. An intense burst of gamma rays washed over a suite of Vela Satellites
77. Around the Pulsar 1829-10, in 1991 by Andrew Lyne

78. The Milky Way and Andromeda galaxies now rushing towards each other are expected to merge in about 4 to 5 billion years. The merged system has been dubbed Milkomeda!
79. The constellation Capricornus
80. Ultra-Compact Dwarf Galaxies. M32, NGC 205
81. Italian-Dutch satellite, which first detected the fading X-ray afterglow of gamma ray bursts on February 28 1997 from GRB 970228, proving their nature.
82. Tom Boles. Made from his observatory in Suffolk using a 35 cm Schmidt-Cassegrain telescope
83. IC 10 in Cassioepeia. It is only 2.3 million light years away
84. Compton Gamma Ray Observatory. Launched on April 5 1991. Was brought down two years ago after its gyros malfunctioned
85. It is a variable nebula in Cepheus, associated with the star PV Cephei. It faded a lot in September, but few weeks later started brightening up
86. Tom Boles of U K
87. Five
88. In 2132 and 2262
89. Edwin Brant Frost in 1908
90. Orren C Mohler. Director of McMath-Hulbert Observatory
91. John Scoville Hall, Director of Lowell Observatory
92. Philibert Jaques Melotte

93. King of Poland
94. By Jane Taylor in 1806 when she wrote 'The Star' for Rhymes for the Nursery
95. Centre for Observational Astronomy, a well known centre for astronomy on the beaches of Algarvae in Portugal
96. Meeting on Asteroids and Comets in Europe in 2006
97. Merlin Medal awarded by British Astronomical Association for planetary astronomy. Hans-Joerg Mettig received it for 2007, for analysis of Jupiter observations
98. Venus
99. V1316 Cygni is a cataclysmic variable belonging to VGSU class of dwarf novae. On June 10th 2006 it underwent a super outburst
100. Omega Centauri
101. Kleopatra
102. Peter Birtwhistle
103. Centaurus A (NGC 5128)
104. 47 Tucane
105. Globular clusters depending on how densely stars are packed; one being the densest. For example M22 class 7 cluster XII is most open
106. Peter Birtwhistle
107. M22; it was discovered by IRAS satellite in 1985
108. Four
109. Wide angle search for planets
110. Great Red Spot

111. Given by British Astronomical Association for developing light sources which reduce light pollution given to Zeta Solar in November 2007
112. Leo Minoris (LMi); it is a type of dwarf supernova
113. L Zacs in April 1980 from the observatory at Riga, Latvia
114. Wolf Rayet star HD 192163
115. It is a PN in Andromeda, NGC 7662
116. It is a PN in Cygnus
117. Cheeseburger Nebula
118. Siberia, Gulf of Obskaja
119. Kinau Crater after C A Kinau
120. Sir George Airy
121. John C Brown
122. Lord Martin Rees; Sir Arnold Wolfendals
123. The moon; discovered by A C Carrieu on 7th September 1985
124. It is an award given by British Astronomical Association for person with marked ability to make astronomical instruments
125. Peter Wise for new concept in telescope optics
126. They are planetary nebula found in the globular cluster Palomor 6 and NGC 6441, respectively
127. It is a PN in M15
128. William Henry Mahoney Christie (1845-1932); Astronomer Royal from 1881
129. These are features present on the moon
130. White dwarf binary

131. Nearest Neutron star. It emits only X-rays and gamma rays. So not detected for long time. The name means 'not there'.

132. Isaac Barrow; Stephen Hawing

133. A possible binary companion to sun with a long period of 76 million years, thought responsible for the mass extinction

134. HATNET: a network of small telescopes mainly for detecting extra-solar planets; HAT-P-76: a planet detected by this network. This is very close to its parent star, one-tenth the distance of Mercury from sun, hence a searing hot surface of 12500^0C

135. It is an extra-solar planet (hot Jupiter) 63 light years away in Vulpecula Constellation, discovered on 5th of October 2005. it was the first extra-solar planet to be mapped.

136. A white dwarf; Palomar Green

137. Near IR Camera and Multi-Object Spectrometer on the Hubble Space Telescope

138. Near Earth Objects

139. Small freshwater lake in Siberia near the river Tunguska, thought to be the crater created during the Tunguska event of 1908

140. Meteorite that fell in Siberia in 1947

141. M13, Globular Cluster in 1974

142. Ida; an asteroid

143. A patch of sky lying roughly between the pointer stars of the big dipper. It is almost free from absorption by natural hydrogen

144. A meteorite

145. Small area of the sky in Sagittarius, relatively free from obscuring dust, so an optical telescope can 'see' central region of our galaxy and beyond
146. Voyage to Laputa the flying island. The prediction was about Mars having two moons
147. On the Moon
148. About 1.5 degrees. Lower areas (valleys) would be permanently in darkness, so ice can accumulate
149. Trans-Neptunian Objects and Kuiper Belt Objects
150. Minor planets orbiting between Saturn and Neptune; Kowal's object, Chiron
151. In the asteroid belt
152. In Quasars. The luminosity is anti-correlated to the width of the spectral line
153. The Japanese spacecraft Hayabusa was approaching the asteroid Itokowa, finally landing on it
154. Triton, satellite of Neptune
155. Ceres, the first asteroid by Piazzi
156. Selenium and Tellurium respectively
157. Neptune (Neptunium) and Pluto (Plutonium)
158. Venus and Mars respectively
159. Another name for TNO's (verbal play on QB, designated for first such objects)
160. Second closest satellite of Neptune
161. Yes, due to earth's tidal drag. About 3 cm per year
162. Clyde Tombaugh in 1930

163. Proposed by Le Verrier to account for the anomaly in Mercury's orbit (precession of the perihelion)
164. 61 Cygni by Bessel
165. Alvin Clark in 1862
166. By Baade and Zwicky in 1933
167. Fritz Zwicky in 1930 in galaxy clusters
168. Jocelyn Bell and Anthony Hewish in 1967
169. John Wheeler in 1968
170. S2 orbits the black hole in the Milky Way centre with velocity of several thousand kilometres per second. Its orbit enabled the estimate of mass of the black hole as 3 million solar mass
171. It has one tenth the mass of the sun and about 10^{-4} of sun's luminosity
172. Vesta
173. Olbers
174. Marooned off Vesta!
175. Ceres, Pallas, Juno and Vesta
176. Near the centre of our galaxy
177. Bright star cluster in LMC (Large Magellanic Cloud)
178. 109
179. It is an exceptionally luminous (>10^5 solar) star
180. Eta Carina
181. Rapidly rotating A stars with peculiar composition
182. Unit of radio luminosity flux
183. Coronal Mass Ejection, Transition Region and Coronal Explorer; Sun
184. Miranda, Ariel, Prospero, Caliban, Sycorax

185. Venus, 243 days rotational period and 225 days orbital period
186. Mercury
187. Cube of the distance
188. 16 seconds in a million years
189. 3 Tera Watts!
190. Ganymede
191. Saturn; one face of it is much darker than the other
192. Water vapour ejected from hot springs on the surface were detected
193. Kant and developed by Laplace
194. Separation distance at which tidal force of primary object exceeds self gravity of secondary which therefore breaks up! For earth-moon system, it is 20,000 km. For objects of same density, the Roche limit is 2.5 times the radius of the primary
195. Giotto
196. Giotto, Japan's Planet A
197. Luna 3, October 5th 1958
198. Thomas Gold
199. A Fall of Moondust
200. Luna 9, on February 3rd 1966
201. The Steady State Theory of the Universe and that pulsars are powered by a rapidly rotating neutron star
202. Charon, Hydra and Nix
203. Band of hot bright stars (types O and B) forming a circle around the sky, representing the local structure of young stars and

interstellar matter. Belt is tilted at about 16^0 to galactic plane

204. Eugene Andrew Cernan and Harrison Hagan Schmitt

205. December 1972, Apollo 17

206. About 25 times

207. About 1 watt per square metre

208. About 60 times solar mass

209. Less than water!

210. An accreting white dwarf which becomes more massive than 1.4 solar mass

211. It is a search for transiting habitable super-earths around nearby M-dwarfs

212. An exo-planet. Very low density (0.3 that of water!)

213. French astronomer Charbonneau

214. First extra-solar planet in whose atmosphere methane and water vapour has been detected

215. Canada-France-Brown Dwarf Survey

216. Brown Dwarfs

217. A variable star

218. The Pioneer Spacecraft launched in 1973 to Jupiter and which has now left the solar system, experienced some additional acceleration in their motion, which is not accounted for even after correcting for all the perturbations (gravitational and non-gravitational)

219. Visible and IR Survey Telescope for Astronomy

220. National Optical Astronomical Observatory

221. AAU Sat 2 (1 kg satellite), carried tiny 200 gram gamma ray detector

222. In the lunar south-west. Its middle is bright and its ends are dark lava plains
223. Occultation between the Uranian satellites Miranda and Oberon, duration of 865 seconds
224. Henry Draper
225. After his death his widow Mary Draper donated the funds that made possible the HD catalogue
226. Morgan-Keenan
227. Morgan-Keenan-Kellman
228. A Jesuit priest who gave designations to lunar Maria, Latin for sea
229. On the moon. It is filled to its rim with dust, dark halo crater
230. Seven and a half minutes
231. Sea of Tranquillity
232. It is a giant elliptical galaxy; 3 billion solar mass
233. Virgo cluster
234. Irvine-Michigan-Brookhaven Detector. Detected ten neutrinos from SN 1987A
235. When a massive star collapses after its inner core is converted into iron, Type IIb
236. Nickel and Cobalt. Ni^{56} à Co^{56} à Fe^{56}, about one solar mass
237. Massive Wolf-Rayet stars
238. Kepler
239. Type Ia
240. 1006 A. D.
241. They are flares associated waves observed to propagate across solar disc at speeds between 500-1500 km/s (visible in UV and H-alpha)
242. American astronomer Gail Moreton in 1960

243. Peter Birtwhistle 100th discovery of main belt asteroid 2008 GE_3 on April 7th 2008
244. Two other main belt asteroids 2008 GB_2 and 2008 GD_3
245. Ulysses Spacecraft
246. Analysis of interstellar medium of isolated galaxies (far IR and radio survey)
247. Surface brightness fluctuations, elliptical galaxies
248. It is a merging galaxy cluster
249. Compact galaxies
250. Hermes, 1937 UB
251. Karl Reinmuth of Heidelberg
252. It was 'lost' for several years and rediscover in October 2003
253. Trans-Neptunian object, every three orbits Neptune makes, Plutino makes two orbits
254. On the sun
255. Well over a hundred!
256. First photographic discovery of an asteroid by Max Wolf in 1891
257. 323 Brucia
258. 90377 Sedna in 2003
259. Hekate in 1868
260. Astronautica in 2000
261. Discovered in 2007 by Dutch school teacher Hanny Van Arkel, who was a volunteer for the galaxy zoo project; it's a small galaxy or dust cloud close to the galaxy IC2497
262. IC2497
263. Perchlorate

264. Lowell Observatory Near Earth Object Search
265. It is an online astronomy project to classify over a million galaxies
266. Asteroid 4522 Britastra discovered by Ted Bowell to commemorate british Astronomical Association centenary in 1990
267. TYC 1343-1865-1
268. Between half and full moon, when moon is 3/4
269. It is a Kuiper Belt Object, it's a large Plutino
270. Asteroid number ten thousand discovered in 1951
271. First asteroid found with a very elliptical orbit
272. About seven degrees
273. Pluto!
274. 2003EL61
275. 68
276. Near Earth Object Confirmation Page; objects, from their orbits, that are near earth appear on this page
277. Optical Gravitational Lens Experiment; to detect compact objects in galactic halos
278. Whirlpool Galaxy
279. Carbon stars
280. About one billion degrees
281. Steven Weinberg
282. This was the period in the early universe when the light elements, D, He-3, He-4 and Li-7 were synthesised
283. Five hundred million photons and less than one proton!
284. Homer; The Odyssey

285. Phoenix Mars Lander in June 2008
286. Ulysses. About 18 years studying the sun
287. It was 'out of ecliptic' and passed over the poles of the sun
288. Fell in Tanzania in 1938
289. A double star, part of the Plough group of stars of Ursa Major
290. Titan and Enceladus mission, planned by ESA to the Saturn system to explore Titan and Enceladus, both strong candidates for possible biological life
291. 1978; 19 years
292. H atom, has one electron
293. $n = 17$
294. Absolute luminosity of late type FGK stars
295. n à $n + 2$
296. Maser emission in radio frequency
297. Mercury Surface, Space Environment, Geochemistry and Ranging
298. October 2008, September 2009
299. Lenticular shape (E7) and Spiral dim glow bright centre (Sd)
300. NGC 4526 in Virgo (E7) and NGC 1494 (Sd)
301. Mars 1999; defective valve spoilt the mission
302. International team transmitted series of radio signals in the direction of four sun-like stars 50-70 light-years away using 150 kilowatt transmitter coupled to 70 m radio-telescope in Ukraine (for SETI)
303. M87 jet, 2 kpc long; with Hubble telescope in 1998

304. 11 billion light-years
305. Ring galaxy, resulting from head on collision with high speed galaxy passing through its centre
306. Low surface brightness
307. UKS 1927-177
308. January 3
309. Gunn-Peterson
310. Multi-aperture red spectrometer on Kitt Peak 4 m telescope
311. J1623, high redshift quasar, $z = 6.3$
312. Bright T dwarf, only 3.6 pc from the sun; Scholz et al in 2003
313. 30 Jupiter mass
314. Gemini South 8 m
315. The seven sisters
316. Saturn's A and B rings
317. Wolf Rayet stars in 4 II region in LMC
318. In SMC, where neutron stars triggered X-ray outbursts stripping material off stellar companions
319. Rossi X-ray Timing Explorer
320. Io, Jupiter's moon
321. Io
322. M105
323. Earth sized features shaped like tadpoles observed in the sun's atmosphere (first observed on 21st April 2002 after a huge flare)
324. 2 Micron All Sky Survey
325. 1.3 m at Whipple Observatory in Arizona and later similar 1.3 m scope in Chile
326. In June 1997; mapped about five million images containing nearly half billion objects

327. On Mars, area with subsoil water expected
328. Abell 160, galactic cluster
329. First tracking and data relay satellite, deployed in April 1983 from space shuttle
330. A white dwarf and red dwarf component stars
331. 7500 years
332. Hare and Peacock
333. 4 million years
334. Heavy isotope, half life of 1.5 million years, produced in supernovae
335. In the spherical halo
336. 12-13.3 billion years
337. Thin disc is younger. Oldest stars in the thick disc are 3-5 billion years older than those in the thin disc.
338. Colour-Magnitude diagram
339. The Japanese Space Agency
340. The two surface rovers on Mars
341. 50 years
342. 14 days
343. Calcium-Aluminium rich
344. In meteorites such as Carbanaceous Chondrites
345. Carbanaceous Chondrites
346. Pu-244; half life is 80 million years
347. About 8 billion years
348. Seems a symbiosis between X-ray AGN and a star burst galaxy
349. Intermediate mass black hole (IMBH) of ~1000 solar mass
350. Krypton
351. 15 times that of earth

352. California Extremely Large Telescope, 30 m planned
353. Giant Segmented Mirror Telescope (U S funded)
354. Seven
355. It is a very young Planetary Nebula, a double shell, one with jet like components, also VLA observations made at 7 mm
356. The first one discovered in QB (cubee)
357. The Atacama Large Millimetre Array, in Chile
358. 200 million years after big bang
359. About 20,000
360. Cloudcroft, New Mexico
361. Mercury
362. Located in the Pistol Nebula near galactic centre
363. At least five million times solar
364. It is a Luminous Blue Variable (numbers refer to co-ordinates) perhaps the most massive (150 solar mass) and luminous star (20 million solar)
365. Stephen Eikenberry at an ASS meeting
366. American Astronomical Society
367. Radiation Pressure
368. It is located fifty thousand light years away on the far side of Milky Way and intervening dust obscures it
369. Eta Carinae
370. It is a spiral galaxy that is plunging through the middle of galaxy cluster Abel 2125 at 2000 km/s. Ram pressure from hot cluster gas is

stripping away this galaxy's gas, till it loses all its gas and is no longer a spiral

371. Why the most massive galaxy clusters today contain few spirals
372. This was earlier thought to be a supermassive star in LMC (of thousand solar masses) but now known to be a collection of several O stars
373. The International Gamma-Ray Astronomy Lab
374. The European Space Agency (ESA) in 2002 October 17th
375. Four telescopes for multi-wavelength astronomy
376. Tens of KeV to several MeV gamma rays
377. High Energy Astronomy Observatory
378. NASA's SAS-2 in 1973
379. The French-Soviet mission for gamma ray imaging
380. The Al-26 decay line, lifetime of a million years; traces star formation
381. Those corresponding to decay of Ni-56 and Co-56
382. Gamma ray satellite launched by ESA in the 1970's
383. Titanium-44, half life of 90 years
384. 2004 JG_6, 6 month orbit, only Mercury is closer
385. 61 Cygni, by Bessel in 1838
386. 1842
387. Immanuel Kant, in 1755, Laplace later developed the theory
388. George Lamaitre in 1929
389. Alexander Friedmann in 1924

390. John Mather and George Smoot for precision measure of the cosmic microwave background using COBE satellite
391. Anthony Hewish in 1974
392. In 1967, for elaborating on the nuclear reactions which generate energy in the sun and other stars, especially pp and CNO cycle
393. Vesto Slipher in 1920
394. Bunsen and Kirchhoff in 1859
395. 1/50,000
396. Lithium
397. About a billionth
398. About one-fourth
399. Beta Librae
400. It is the only solitary star reported to appear green
401. It is the SETI institute project for detecting ET radio signal
402. From hundred light years, for transmitter of ten kilowatt and 300 m diameter antenna
403. Star with extremely eccentric orbit, which comes once in 50 years to within eight light hours of the galactic centre
404. Formation of Titanium oxide and other oxides
405. U Scorpii, close to the limiting mass
406. A bright Mira variable, with a large brightness range
407. Sakurai's object in the constellation of Sagittarius, discovered in 1996, is believed to have undergone Helium flash to become White Dwarf (so called third dredge up phase)

408. Nine
409. Satellites to detect X-rays
410. X-rays blast from a magnetar
411. In 2025
412. Tiny dust particles at a distance of thousand astronomical units from the star
413. Terzan 5, 100 metre Green Bank Radio Telescope
414. Ultra high energy cosmic rays
415. Quasar
416. Riccardo Giacconi in 1962
417. Raymond Davis
418. Riccardo Giacconi and Toshiba from Japan
419. On Io
420. Galileo spacecraft in August 2001
421. In 1986, brought down to a fiery recently on March 23rd 2001
422. On April 7th 2001
423. The r or the rapid neutron capture process, producing the heaviest elements
424. It is a 3.5 m telescope atop Kitt Peak in Arizona
425. On Io
426. The Sombrero Galaxy
427. The Beehive Cluster
428. A gorgeous double star at head of Cygnus
429. M33
430. Largest sunspot group found in March 2001, led to a large flare on April 2nd 2001
431. X20
432. Nothing to do with the moon! Molybdenum Observatory of Neutrinos uses the Molybdenum

isotope Mo-100, which captures a solar neutrino to become Tc-100, has a low energy threshold

433. High Amplitude Delta Scuti stars. They are in the process of evolving off main sequence stars, pulsating variably.
434. Evolved population II HADS
435. A population I HADS variable star
436. Tsesevich in 1939
437. Chris Graham Remote Telescope in Pingelly, Western Australia
438. Canadian Satellite Tracking and Orbit Research
439. About 2050 satellites from January to December 2007
440. The Dogon tribe in Mali in Africa had apparently a hoary tradition that Sirius had a very dense companion star made of Sogulu, a metal so heavy that one grain of it weigh as much as a donkey load
441. It is a 60 inch telescope in the Boyden Observatory
442. In 1889, near Lima (South America) about fifteen kilometres from small town of Chosica. In 1890 it was moved to Arequipa in Peru
443. An engineer Uriah Boyden, in Boston left 238 thousand dollars to Harvard College for extending astronomical research
444. Edward Pickering, director of Harvard College observatory

445. Since it was a joint venture by Armagh, Dunsink and Harvard Observatories

446. It is light back scattering caused by back scattered light reflected from clouds of water droplets

447. Instituted by Augustus Chant, a Canadian astronomer

448. Given to a Canadian amateur astronomer for original investigation; Geoff Gaherty received it in 2008

449. Topham in 1940

450. Wilkinson Microwave Anisotropy Probe, a satellite launched in 2003 to measure the anisotropy in the cosmic microwave background

451. In 2008, to measure cosmic microwave background anisotropies on a small angular scale

452. He shared the prize in 1974 with Anthony Hewish. It was for his work on 'aperture synthesis'

453. It was a balloon borne experiment launched from the South Pole in 2001 to study the cosmic microwave background

454. A supernova explosion caused by collapse of a very massive star, releasing more energy than an ordinary supernova, and also followed by Qr preceding a gamma ray burst

455. A white dwarf accreting matter from a giant or main sequence star. The matter piles up and after getting heated up to several million degrees undergoes a nuclear explosion ejecting the accreted matter

456. Fox
457. It is a low mass low (0.1 solar mass) luminosity Red Dwarf
458. A yellow giant more evolved than the sun
459. White Dwarf
460. A blue giant; B spectral type about 15 solar mass
461. About 160,000 light years away in the Large Magellanic Cloud (LMC)
462. Stephenson-Sanduleak
463. Venus
464. In second and third week (16th onwards) of July 1994
465. Three kilometres, the largest fragment
466. About six million megatons of TNT equivalent
467. It is equivalent to one Hiroshima bomb exploding every second for a period of ten years continuously
468. About sixty km/s, the escape velocity of Jupiter
469. These are dark nebulae where stars are beginning to form, found by Bart Bok
470. 13.2 billion years old
471. From its very low iron abundance compared to sun
472. The Dumbbell Nebula, M27
473. Charles Messier on 12th July 1864
474. Vulpecula
475. Wild Duck Cluster
476. William Henry in 1844
477. Max Planck Millimetre Bolometer-2

478. Large Apex Bolometer Camera
479. Atacama Pathfinder Experiment, between ESO and Onsala Space Observatories
480. The sudden change in density between Crust and Mantle in the earth's interior
481. Between Mantle and liquid Core of the earth
482. Full Width at Half Maximum and Point Spread Function
483. Perseus, Scutum-Centaurus, Sagittarius and Norma
484. No, it is increasing at about 3 cm a year
485. Laser corner reflectors left behind on the moon by the Apollo astronauts have enabled estimate to an accuracy of less than a centimetre of the moon's distance.
486. The moon's tidal friction is slowing down earth's rotation and hence from conservation of angular momentum implies that the distance increases
487. It is expected to increase till the earth's daily rotation is slowed down to about 48 days
488. Barnard 86
489. Epsilon Aurigae with a period of 27 years
490. Plutoids
491. 2008 HJ, rotates once every 43 seconds
492. Richard Miles in April 2008
493. G 1.9 + 0.3, youngest known remnant, expanding at 15,000 km/s, perhaps exploded 150 years ago, not seen then because of heavy obstruction by interstellar dust
494. E V Lacertae, only 16 light years from earth

495. Only four days (fast rotator), field is hundred times solar
496. Geminga
497. 68 degrees North
498. Dodo
499. Three earth mass
500. On Mars
501. The Mars Opportunity Rover had one year ago, got into it to study ot and now has climbed out
502. Bank of river Orwell in Suffolk
503. Auroral Kilometric Radiation
504. Great Observatory All Sky Luminous Infrared Galaxy Survey
505. It is devoid of dust
506. 'Horned Goat'
507. A large number of 'water-based' constellations are located there
508. Tanzania in 1938
509. On the moon, close to the giant Apenine mountains
510. Geminiano Montanari in 1669
511. Astronomy Common Object Model
512. On Mars, with a length of ten thousand kilometres. It is perhaps the largest impact crated in the solar system
513. The Iridium Satellites, launched by Motorola Company, for communication purposes are large flat mirrors in orbit. When illuminated by sunlight, they can sometimes shine (very briefly) at 8.0 magnitude.

514. It is presumably a new mineral (containing a new Manganese Silicide) found inside an interplanetary dust particle
515. New mineral has been named in honour of Donald Brownlee, an expert on interplanetary dust.
516. It is thought to have come from comet 26P/Grigg-Skjellerup
517. 1902
518. Nebensonneu
519. Wilhelm Muller, "Three suns I saw stand in the sky"
520. Great Observatories Origins Deep Sky Survey
521. On Venus
522. Volcanic regions on Mars
523. Latin, meaning ditch or trench
524. On Mars, the Phoenix Spacecraft was 'digging' it
525. A Flemish Jesuit astronomer who served the Kangxi Emperor in China around 1673
526. He designed and made a set of six new instruments for Beijing's observatory tower in 1699
527. Matteo Ricci and Adam Schall Von Bell
528. A great bronze celestial globe, 2 metres in diameter with a claimed cost of fifty thousand silver pieces
529. Utagawa Kuniyoshi in 1827
530. As a military strategist with his knowledge of maps, navigation, various technologies, etc. The woodcut in Q.539 portrays him as Chitasei Goyo
531. Danish astronomer, Tycho Brahe
532. On Mars

533. Phobos, satellite of Mars
534. Mars Orbiter Laser Altimeter
535. The Rosat in 1991 when it found that its X-ray emissions pulsate every 0.237 seconds
536. A type F supergiant
537. The red dwarf Eliese 436
538. Thirty
539. Its surface does not vibrate, unlike alpha centauri A or the sun
540. About 8 percent
541. It is a double star in the open cluster NGC 6568, discovered by S Aravamudan of Nizamiah Observatory
542. Victor Hess
543. Ganymede
544. 5265 km as compared to 5150 km of Titan, so just about 100 km larger!
545. Themisto
546. A large dark area, 3000 km across on Ganymede
547. Galle ring
548. Galaxy Evolution Explorer, maps the UV radiation from various sources
549. Soviet-American Gallium Experiment, using several tons of gallium to detect the low energy solar neutrinos from the pp reactions in the solar core
550. Galatea
551. About 0.1 nano Tesla
552. 1612 MHz
553. Water masers, methanol masers, SiO masers, etc

554. Araki-Alcock Comet, found by IRAS satellite and independently by G Araki and George Alcock
555. Infrared Astronomical satellite
556. IRIS
557. Iocaste
558. La Palma in Canary Island
559. The ING (Isaac Newton Group) jointly owned by UK, Netherlands and Spain
560. The Jacobus Kapteyn Telescope also part of ING
561. International Sun-Earth Explorer
562. Inter-galactic medium
563. Epsilon Bootis, consisting of KO giant and AO dwarf and visually appearing to have lovely colours, orange plus blue-green, etc. The word means 'most beautiful'
564. 3.5 m New Technology Telescope at La Silla
565. In the Atacam desert Chile
566. Lepus, in the Southern Hemisphere
567. John Herschel
568. Josep Sola in 1907
569. Gerard Kuiper in 1944
570. More than 5 times
571. About 16 earth days
572. It was a meter sized object which in September 2006 was temporarily captured by the earth's gravity
573. Three
574. A measure of the number of spots on the sun
575. Royal Greenwich Observatory

576. Around September 23rd, autumnal equinox, full moon rises only 18 minutes later on successive evenings
577. Royal Edinburgh Observatory
578. Ole Roemer in 1676
579. Clusters of galaxies
580. The Royal Astronomical Society
581. Rubin-Ford Effect
582. The Ryle telescope
583. On surfaces of meteorite larger than 10 cm
584. Regulus
585. Newton in 1668
586. Variable stars pulsating in the fundamental mode and first overtone respectively
587. By the r process (rapid neutron capture) in the aftermath of a core collapse type II supernova
588. Mercury
589. Solar Anomalous and Magnetospheric Particle Explorer
590. Produced Carbon, also called Salpeter process
591. Sakigake
592. Schiaparelli
593. Karl Schwarzschild in 1905
594. These are mass concentrations under the moon's surface that disturbed the orbits of the NASA Lunar Orbiter Mission
595. They posed a navigation challenge to the Apollo Spacecraft and would be fatal for low altitude orbiters which do not carry auxiliary propulsion system

Study of the moon's shape

Gravity Recovery and Climate Experiment Spacecraft, which are twin geodetic satellites in closely spaced earth polar orbits. Tiny separations in their distances accurately map deviations of earth's gravity field

598. March 2002
599. One of the outer most tiny moons of Jupiter
600. Mirach, that is beta Andromedes
601. Stands for Gravity Recovery and Interior Laboratory, to be launched in 2011, to study in detail the moon's gravity field. It's two spacecraft would fly just 30 km above lunar surface
602. Mintaka, Delta Orionis
603. The recently deployed Japan's lunar mission's orbiting triad of spacecraft
604. Selenoid
605. Multi-conjugate adaptive optics
606. It uses multiple deformable mirrors phased to laser beams that create 'guide stars' by exciting sodium atoms in different upper atmospheric layers
607. European Extremely Large Telescope
608. 42 meters
609. Nectaris
610. Containing least mare basalt
611. On the moon
612. Mare Imbrium
613. After the goddess Sedna, of Inuit legend who lives below the frigid artic seas. As Sedna's

temperature never rises above 30K, due to its great distance from the sun, it got this name

614. GRACE
615. M31, the Crab Nebula
616. Lord Rosse in 1845
617. The Burst and Transient Source Experiment aboard the Compton Gamma Ray Observatory
618. Australia
619. Zakari'ya Al Kazwini by seeing the star Alcor
620. Mizar is six times brighter than Alcor
621. About ten billion light years
622. Milkomeda, the merged remnants of Milky Way and Andromeda
623. The Peekskill meteorite weighing 12.6 kg, named after the place it fell
624. Alpha Andromeda
625. P-class asteroid, having a featureless reflectance spectrum
626. Mars
627. Philip bands
628. Portia
629. Porrima
630. A pore
631. Solar radiation pressure, acting on micron sized particles
632. The Pribram meteorite, broken up into 19 fragments, photographed by cameras operated by Ondrejov Observatory in Czech Republic
633. Publications of Astronomical Society of the Pacific (PASP)
634. Q-class asteroid

635. Quasi-stellar radio source
636. On Mars
637. Radio Telescope in southern Russia
638. Dembowska
639. The Saturn Nebula
640. A reflecting telescope that incorporates three spheroid concave mirrors. Derived from German for 'tilted mirror'
641. S-class asteroids, S stands for Silicaceous
642. Schmidt-Cassegrain telescope
643. Space Telescope Science Institute founded in 1981 at Johns Hopkins University
644. Suisei, observed Halley Comet's hydrogen halo in 1986
645. Surge prominence
646. Sycorax
647. Syndyne
648. Syrtis Major Planum
649. Technetium
650. Promethium-61
651. Spallation
652. Speculum metal
653. Spinars
654. Toby Jug Nebula
655. Toro
656. Triangulum galaxy
657. Open stellar cluster
658. Tully Fisher relation
659. Faber Jackson relation
660. Different types of galaxies in the Hubble classification

661. Tuttle-Giacobini-Kresak, named after the three astronomers involved
662. Tycho catalogue
663. The water jar (in Aquarius)
664. It is the Wide Field and Planetary Camera 2, with corrective optics, installed on the Hubble orbiting telescope during servicing mission 1 in December 1993
665. This was in March 2002, when Columbia astronauts installed the Advanced Camera for Surveys (ACS) and new cooling System for NICMOS, on the Hubble telescope
666. Near Infrared Camera and Multi-Object Spectrograph
667. During servicing mission 2, in February 1997
668. Hubble Space Telescope
669. Betelgeuse
670. Centre for High Angular Resolution Astronomy
671. The star Altair in 2007
672. Spins at one million kilometres per hour
673. Sixty times faster
674. 0.003 arc seconds
675. 572
676. Two!
677. 0.002 percent
678. 5 minutes 10 seconds, mostly from Australia
679. Near Cape Girardeau, Missouri
680. August 2nd 2027, February 16th 2045, May 22nd 2096
681. The Meteoritical Society, M Wadhwa for contributing to meteoritics

682. 83T6 Wadhwa
683. Centre for Meteorite Studies, University of Arizona
684. W Baade, Hidalgo
685. It has a very elliptical orbit, with aphelion beyond Saturn
686. Heaviside layer
687. Helene
688. Hanle effect
689. 10^{-2} to 10^{-3} Tesla
690. 5 KeV to 100 KeV
691. Harvard College Observatory; in 1839
692. 2061
693. Hakucho in February 1979
694. 'Far away' in Japanese, known as Muses-B before launch and then HALCA which stands for Highly Advanced Lab for Communication and Astronomy
695. Pope Gregory XIII in October 1582
696. Green Bank Telescope
697. Hadar
698. Hadley Rille
699. Neptune, found by Voyager 2 in 1989
700. Comet Grigg-Skjellerup in 1992
701. Grotrian diagram
702. Gnomon
703. Ginga
704. Giant Molecular Cloud
705. GRANAT
706. Goddard Space Flight Centre in 1959
707. Gran Telescopia Canarias, a 10.4 m reflector at an altitude of 2.5 km in Canary Island

708. James Gregory
709. Red Giant variable star, discovered by John Hind who noted its blue-red colour. It is also called Hind's Crimson Star
710. Hoba West meteorite found at Hoba farm, Namibia, with estimated mass of sixty tons
711. The Homonculus Nebula
712. Jeremiah Horrocks
713. On Mars, example of horst, that is a strip of land uplifted between parallel faults
714. Hohmann ellipse
715. Homberg radius, measures size
716. HZ43
717. International Year of the Quiet Sun
718. One sansa Watt, that is 10^{30} Watts
719. Melipal
720. Southern Cross in the local Mapuche language
721. F Whipple and D Menzel
722. The average density of the sun which he stated is just above that of water.
723. Mimas, moon of Saturn
724. Henry Draper
725. Damjon scale (running from 0, the darkest to 4, very bright)
726. Davida, with diameter of 325 km
727. Dawn, to travel to Vesta
728. Despina, inside Le Verrier ring
729. Dustless
730. Supergiant ellipticals at centre of rich clusters of galaxies
731. Bianchi cosmology

732. The Big Crunch
733. The Hubble constant, it scales as the square of this constant
734. About 20 hydrogen atoms per cubic metre of space
735. Bepi Colombo
736. Bellatrix
737. Jocelyn Bell
738. CP1919
739. Barnard's loop
740. Mars
741. Potentially Hazardous Asteroids
742. About 900
743. They were found to be the same object
744. 12 million km on March 25th 1996
745. Johannes Hevelius
746. Mithradates VI, king of Pontus in 119 BCE
747. Nero
748. Mawangdui Silk Book from China
749. A very luminous cluster in the Milky Way, a grand star forming region, which in one cubic light year hosts wide variety of hot massive stars
750. The moon
751. Beehive star cluster, as it is crowded with stars
752. Hebe, by Prussian Karl Henche in 1847
753. Hebe was the daughter of Zeus and Hera in Greek Mythology, embodiment of youth who poured the nectar at the Olympic table
754. Athabasca Valles
755. A 2.5 degree long string or chain of stars
756. Eratosthenes around 230 BCE

757. Hydrogen, helium, oxygen, carbon, iron

758. It is an example of a gravitational lens effect in which four images of a background object are formed, arranged in the form of a cross (eg. Huchra Lens)

759. Frank Dyson. In 1919 he led the eclipse expedition which verified Einstein's prediction of the deflection of light by the sun's gravity

760. White dwarfs with a continuous featureless spectrum

761. Showing only broad absorption lines of hydrogen, eg. Sirius B

762. On Mercury. It is a ridge on a planetary surface

763. An M type red dwarf whose spectrum has emission lines

764. John Dollond, in 1753 and 1757 respectively

765. Dirty Snowball Model

766. A radio receiver to measure very weak signals in the presence of noise

767. On Venus, it is a deep trough (*1000 km* X *100 km*) in the middle of Aphrodite Terra

768. Diffuse interstellar bands, broad absorption features in distant stellar spectra caused by interstellar material

769. Mars, now called Copratos, a dark elongated feature

770. B Lyot in 1930

771. Clown Face Nebula, NGC 2392

772. January 1995, Geographos

773. It used up all the control gas

774. Eris A, dwarf planet, earlier called Xena
775. 31 Euphrosyne, one of the larger asteroids, 250 km diameter, discovered by James Ferguson in 1854
776. They are satellites of Uranus and play the role of 'shepherds' maintaining the Epsilon ring of the planet
777. A gap in the Saturn A ring
778. It is one of the darkest of asteroids with an albedo of only a few percent. Discovered in 1867 by Robert Luthor
779. The pilot of the fleet of king Menelaos
780. They are small outer moons of Jupiter
781. They are satellites orbiting Saturn in the same orbit
782. It is a narrow gap towards outer edge of bright A ring of Saturn
783. 10 Hygeia, discovered in 1849 by de Gasperis
784. It is an extended source of infrared radiation in the Orion Nebula
785. Sir William Huggins, in 1868, first measured the Doppler shift in the spectrum of Sirius
786. Hamal, the brightest star in Aries
787. It is a large solar observatory with a 1.6 m mirror, located at Kitt Peak, completed in 1962. produces a high resolution image of the sun, 30 inches in diameter
788. It is a four metre optical reflector telescope at Kitt Peak, operating since 1973
789. Sir Norman Lockyer
790. Fomalhaut

791. Lincoln Near Earth Asteroid Search

792. LaCerta, small constellation between Cygnus and Andromeda

793. Johann Hevelius

794. For her discovery of the relation between the period variation in brightness and the absolute luminosity of Cepheid stars

795. The open cluster NGC 3324, popularly known a the Jewel Box

796. The group of three stars, epsilon, zeta and eta in Auriga. Comes from the name of alpha Auriga named Capella, the little she goat

797. Percival Lowell, 24 inch refractor

798. Kennedy Space Centre

799. Small outer moon of Uranus

800. Small outer moon of Saturn, 14 km in diameter, found in 2000, has a very elliptical orbit

801. Group of asteroids with orbital planes inclined at 24 degrees

802. Minkowski's footprint

803. Delivered by Luna 21, on January 16th 1973 at Mare Serenity. In four months this eight wheeled lunar rover covered 37 km distance

804. The Large Binocular Telescope, twin 6.84 m telescope located at Mt. Graham Arizona, equivalent to a single 11.8 m mirror

805. Kuiper Airborne Observatory

806. Discovered by Palisa in 1880. Radar observations revealed a very unusual shape, called 'dog bone', composition considered to be very metallic

807. They are located at L4 and L5, 60^0 on either side in thesame orbit as Jupiter around the sun

808. As long as ratio of masses of the two large bodies (in the restricted 3-body problem) exceeds 25

809. The Carrington rotation number sequence, rotation of sunspots

810. The Robert C Byrd Green Bank Telescope

811. The Sagittarius Dwarf Galaxy

812. In Sculptor

813. About 220 km/s

814. Infrared Optical Telescope Array consisting of three 0.45 m collectors

815. John Franklin-Adams

816. Erinome

817. Equinoctial Colure

818. Achernar and Acamar respectively

819. Equuleus

820. The equivalent width

821. The Faulkes telescope, a pair of 2m reflectors for use by students in U.K, Australia, etc. over the internet

822. Fibrils

823. The Crepe ring

824. Crimean Astrophysical Observatory

825. The C line

826. When latitude of sunspots are plotted against time

827. Benjamin Boss

828. Planetary Nebula

829. Harold Babcock and Horace Babcock in 1952

830. White noise
831. Brightest star in Cannes Venatici. It is Latin for 'Charles's Heart', the name was given by Charles Scarborough in 1660, to mark the execution of Charles I
832. Achernar, equatorial diameter sixty percent larger than polar diameter, rotating at 250 km/s
833. Arcturus, a population II star, large proper motion
834. It is the observatory of the University of Toronto, presented to the university by Mrs. Dunlap in memory of her husband
835. A 1.9m reflector, the largest in Canada
836. Henry Draper, whose catalogue of stellar spectra has given the prefix HD before several objects
837. Henry Draper in 1880
838. In the Tenerife Island, also has a solar observatory
839. Small outer moons of Jupiter
840. The former is a bright star while the latter is a small outer moon of Jupiter
841. Donati's comet
842. Schedar or Alpha Cassiopeia
843. Scheat or Beta Pegasi, a super giant M star in Pegasus
844. Kitt Peak operated by the universities of Wisconsin, Indiana, Yale and National Optical Astronomical Observatories

845. Change in sunspot appearance due to solar rotation. The penumbra nearest the limb appears wider.

846. One of the largest Planetary Nebula known, M97 in Ursa Major

847. Moons of Saturn

848. Alternative name for the star Alpheratz

849. Near St. Petersburg in Russia, in 1718

850. A large lunar crater, NW of moon. It has high walls and a central peak

851. Supernova remnants showing no shell structure; Crab Nebula

852. It is the 1.2 m Schmidt Telescope at Mt. Palomar

853. It is a working model of solar system showing planets and moons in orbit around sun

854. The term comes form a model made in 1713 for the Irish nobleman who was the fourth Earl of Cork and Orrery

855. 511 Davida

856. S stars

857. 719 Albert, after Baron Albert Rothschild

858. Launched in 1992, collaborative effort by NASA and Japan to explore the magnetotail, one million kilometre behind earth

859. It is an open star cluster in Cassiopeia. It is the second entry in Jurgen Stock's 1956 catalogue of more than twenty open cluster

860. Segin, Epsilon Cas

861. Ruchbah

862. 103

863. Pierre Mechain

864. A meteor that explodes in the earth's atmosphere
865. Study of physical chemical processes in planetary upper atmosphere
866. 85P/Boethin
867. Filipino Rev. Leo Boethin in January 1975
868. The dearth of helium atmosphere white dwarfs in the temperature range 30,000-45,000 degrees.
869. They are planned spaces probes to quantify the dark energy content of the universe. Joint Dark Energy Mission and Supernova Acceleration Probe
870. Advanced Dark Energy Physics Telescope; in 2015
871. Nine from 1917-2008
872. SN 2008 S, Ron Arbour
873. Mars Science Laboratory to be launched in 2009. The Eberswald Crater (an ancient river delta and the Holden Crater)
874. Bradford Robotic Telescope
875. Saturn's rings would appear edge on as seen from earth
876. The south face of the ring would not be illuminated by the sun
877. It is the year of least solar activity since 1954. There have been 205 days without any sunspots as compared to 245 days in 1954
878. Under certain conditions, a temperature inversion can produce an unusual visual mirage,

the apparent rise of the sun long before its forecasted time

879. From a group of Russian Islands in the Artic, where in January 1597, Gerrit de Veer, a carpenter first recorded the mirage

880. Makemake (pronounced Mah-keh Mah-keh)

881. Three including Pluto and Eris

882. The Polynesian creator of humanity and god of fertility

883. It underwent a sudden brightening to a nova phase which proved to be one of the brightest over the past 35 years

884. The Spitzer Space Telescope identified this star at more than three million times the solar luminosity; could be brighter than Eta Carina

885. Mercury

886. Enceladus, moon of Saturn

887. Means 'furrows', complex network of parallel ridges and depressions on the surface

888. A division in Saturn's rings in the outer part of the Crepe ring found in 1980 by Voyager I

889. He explained with a theoretical model from stability considerations that Saturn's rings cannot be continuous but must be made of discreet small objects

890. Chemically Peculiar star

891. Maria family (170) Maria being the largest member

892. Josheph Perrotia in 1877

893. Statistical selection effect in galaxy surveys

894. Manganese star, late B type spectral type

895. Mercury-Manganese star
896. On Europa, Jupiter's moon; it is a dark spot
897. Magnetic inversion line
898. Lyot filter
899. On Mars; landslides
900. K Correction
901. Kellner eyepiece
902. Crystalline rocks from lunar highlands. K for potassium, REE for rare earth elements, P for phosphorus
903. Grism
904. True
905. Tidal force is proportional to the mass by the cube of distance. Sun and moon subtend the same angle of half degree on earth. So ratio of their diameters is same as ratio of their distances. Hence the answer follows
906. Kepler's III laws and that the sun subtends half degree on earth
907. Fermi Gamma Ray Space Telescope
908. It is the Dark Energy Survey which is being carried out at the Anglo-Australian Telescopes
909. Started in August 2006 and ends in July 2010
910. Solar Ultraviolet Magnetograph to measure magnetic fields in solar transition region, by investigating Zeeman splitting of UV emission lines in active regions
911. Typical variation of solar activity with a characteristic time scale of 60-150 years
912. About nine nano lumens per square metre
913. Six photons per second

914. Eye relief

915. Smallest section through light beam from an eyepiece through which all the light from the eye piece passes (lens closest to observer's eye is called eye lens)

916. At higher altitudes in the solar atmosphere, flow (~20 km/hr) is directed both inwards and downwards towards sunspot

917. Moat

918. European VLBI Network

919. Evection

920. E layer of the ionosphere

921. 206,265

922. Galactic year or cosmic year

923. Platonic year

924. A large low plain, a feature on Enceladus, Saturn's moon

925. Period-luminosity-colour relation

926. Shield volcanoes, large area and gentle slope

927. A Catoptric syatem, that is an optical system using only mirrors

928. Searches for asteroids and comets passing close to earth on areas away from ecliptic and at high inclinations. Now called Catalina Sky Survey from 1998; uses 0.4m Schmidt Telescope

929. Nathaniel Bliss, 1762-64

930. Francis Graham Smith

931. Combined Array for Research in Millimetre Wave Astronomy

932. Classification scheme for clusters of galaxies

933. A bimonthly popular magazine published by Astronomical Society of the Pacific
934. The near horizontal path on the HR diagram for stars of different mass evolving to MS
935. Milton Humason
936. Hendrik van de Hulst in 1944
937. Pribram meteorite in 1959, near Prague
938. The Ondrejor Observatory in Czech Republic
939. Spray
940. Z Camelopardalis stars
941. Uranometria
942. Van Maanen's star, 14 light years away
943. Van Biesbroeck's star
944. Vatican Advanced Technology Telescope, 1.8 m
945. Wezen
946. The paucity of stars in the HR diagram between luminous blue variable and red supergiant
947. Faint Irregular Galaxy GMRT survey
948. 120 days
949. One year!
950. About 6 days
951. Two hours
952. About twenty minutes
953. Yes, the impact time is about 0.168 times the orbital period
954. A C Clarke in Jupiter V
955. About 2×10^{21} photons/second or 2 sextillion or 2000 pentillion photons/second
956. 4×10^{10} AU or about 200 kpc
957. About one hundred millionth or 10^{-8}

958. About 10^{13} photons/m^2/s (ten trillion)
959. On the Mars Phoenix Probe
960. Thermal and Evolved Gas Analyser, to record release of gases from heated samples and Microscopic, Electrochemistry and Conductivity Analyser, to measure pH etc. of samples
961. On Mars, it has a 5km high mound of layered deposits
962. On board the Mars Express Orbiter
963. These are planned underground future neutrino telescopes. The first would use a cubic kilometre of seawater as detector. The latter would use either million tons of water or liquid Argon
964. The Einstein Telescope
965. Mars
966. Planetary Nebula
967. The astronomer Henise
968. While climbing Mt. Everest
969. One year!
970. 270 days
971. About on millimetre
972. The Axion
973. Ten billion times lighter than the electron
974. By its resonant conversion into two microwave photons in a cavity permeated by a magnetic field
975. The neutralino, about 100 times the proton mass
976. About 10^{14} m/s^2
977. About a 1000 Gauss (proportional to $1/r^3$)
978. About 10 million tons
979. 6000km/s

980. 0.8 light velocity or 240,000 km/s

981. One of the small satellite galaxies orbiting the Milky Way. Perhaps the most dark matter dominated dwarf galaxy, with a thousand times more dark mass than luminous mass!

982. 490

983. Greisen-Zatsepin-Kuzmin effect refers to the upper limiting cut-off energy of distant high energy cosmic rays (protons) due to interaction with the microwave cosmic background. For protons, the cut-off is about 60 Exa-electron volts

984. The Fred Lawrence Whipple Observatory

985. Absorption by water vapour and oxygen in the earth's atmosphere

986. Mount Stromlo and Siding Spring Observatory and New Technology Telescope

987. Orgueil Meteorite

988. William Parsons, the Third Earl of Ross

989. Forbidden line of doubly ionised oxygen (double bracket) and semi-forbidden line of doubly ionised carbon (single bracket)

990. Two subfamilies of asteroids at a mean distance of 2.4 AU from the sun but with different orbital inclinations

991. Herschel-Rigollet Comet 35P, period of 155 years

992. A dust cloud which is likely the shredded core of a gas giant Jupiter-like planet.

993. Ernesto Capocci in 1850 in a letter to the Belgium Royal Science Academy

994. Robert Wood at Long Island in 1908!
995. Alan Guth when he proposed the inflation model
996. Pulsating white dwarfs with helium atmosphere and temperatures around 25,000 degrees
997. CoW Vir Star with Helium-Carbon-Oxygen atmosphere and temperatures of 120 thousand degrees
998. It was caught in the act of exploding on 9th January 2008, moments after the shock wave blasted through the surface of the blue supergiant progenitor star. X-ray signature detected by Swift satellite
999. NGC 2770
1000. Oscar Wilde in the play 'Lady Windermere's Fun' (1892)
1001. King George V, in 1913 approved this motto for the Royal Air Force. It means 'through struggle to the stars'

□□□